Biology

A* Study Guide for INTERNATIONAL GCSE

Biology

A* Study Guide for INTERNATIONAL GCSE

Pamela Maitland

www.galorepark.co.uk

Published by Galore Park Publishing Ltd
19/21 Sayers Lane, Tenterden, Kent TN30 6BW
www.galorepark.co.uk

Design and typesetting by River Design
Illustrations by Ian Moores

Printed by Charlesworth Press, Wakefield

ISBN 978 1905735 45 7

First published 2012

Details of other Galore Park publications are available at www.galorepark.co.uk.

Acknowledgements

To James, Iain, Laura and De.

The publishers are grateful for permission to use the photographs as follows:

P2/P142 Rene Bhavnani/istockphoto, P3 Nils Kahle, P4 fotografie4you.eu/photos.com, P7 Eye Of Science/Science Photo Library, P9 Rob Stegmann/photos.com, P9 Heiko Kiera/Shutterstock, P10 Bruce MacQueen/Shutterstock, P10 Martin Fasanek/istockphoto, P11 Ali Mufti/Shutterstock, P13 Pakhnyushcha/Shutterstock, P13 Nancy Nehring/istockphoto, P13 Cuiphoto/Shutterstock, P21 Professors P. Motta & T. Naguro/Science Photo Library, P22/P71 Steve Gschmeissner/Science Photo Library, P25 Scott Camazine/Visuals Unlimited, Inc, P26 Martyn f. Chillmaid/Science Photo Library, P26 Cordelia Molloy/Science Photo Library, P31 Evgeny Karandaev/istockphoto, P31 Tom Foxall/istockphoto, P37 J.C. Revy, ISM/Science Photo Library, P37 J.C. Revy, ISM/Science Photo Library, P43 Eos/Wanner/Science Photo Library, P45 Nigel Cattlin/Science Photo Library, P45 Nigel Cattlin/Visuals Unlimited, P53 SCIEPRO/Science Photo Library, P62 Power and Syred/Science Photo Library, P67 Fred Goldstein/Shutterstock, P71 Dr Keith Wheeler/Science Photo Library, P73 Nigel Cattlin, Visuals Unlimited/Science Photo Library, P77 Eric Grave/Science Photo Library, P77 SCIEPRO/Science Photo Library, P78 Martin Oeggerli/Science Photo Library, P101 Solid Web Designs LTD/Shutterstock, P107 Kenneth H. Thomas/Science Photo Library, P109 Roel Smart/photos.com, P109 Melinda Fawver/Shutterstock, P110 Rudkouskiy Yahor/photos.com, P115 Gene Chutka/istockphoto, P117 lusoimages/Shutterstock, P119 NIBSC/Science Photo Library, P126 Mark Burnett/Science Photo Library, P127 Soverign, ISM/Science Photo Library, P130 Look At Sciences/Science Photo Library, P130 Dr. Stanley Flegler/Visuals Unlimited, Inc, P138 Martyn F. Chillmaid/Science Photo Library, P151 Julius Fekete/istockphoto, P153 Dr. Morley Reid/Shutterstock, P154 Petek ARICI/istockphoto, P155 Andrey Kekyalyaynen/Shutterstock, P159 Alan Crawford/istockphoto, P160 Dabjola/Shutterstock, P160 Damian Herde/Shutterstock, P160 Aliaxei Shupeika/photos.com, P167 Amy Tseng/Shutterstock

Contents

Page

Introduction

Biology is a very accessible and relevant science and, when supported by a sound grasp of Chemistry and Physics, a world of wonder and intellectual satisfaction is created. If a student appreciates that all three sciences inform each other, especially when it comes to the approaches and techniques used in investigations, then it will come as second nature to make links between the various topics in Biology itself. The skills of learning facts, concepts and processes and applying them to new situations – for example, the lateral thinking required in so many examination questions! – will earn the top grades and form a sound basis for further study.

What makes this revision guide unique is that it presents the facts as clearly and as logically as possible – but without oversimplification – so that students acquire knowledge **and** understanding alongside the ability to use essential tips which will assist you in achieving an A* grade. The photographs, illustrations, text boxes and examination tips all contribute to making this an essential guide.

As you use this book, you must be interactive: make your own notes either on paper, electronically, by recording yourself, by using index cards, spider diagrams, or post-it notes. Start the year off by trying different ways of studying until you find what works best for you.

Remember Biology is fun, and the more you know about it, the more enjoyable it becomes. If you have chosen this book, you have already decided to earn top grades. Good luck.

How is this book organised?

This book follows the structure laid out in the Edexcel IGCSE Biology specification (4BI0), Edexcel Level 1 / Level 2 Certificate in Biology (KBI0) and, where necessary and appropriate, includes extra material to fulfil the Cambridge International Examinations IGCSE Biology specification (0610).

Material that is only applicable to CIE students is indicated by this symbol: **CAM**

Each section begins with a list of what you are expected to know for that particular topic. There then follows the material that you need to learn. This has been laid out in a revision-friendly way, using bulleted lists and tables to aid visual learning. Worked examples of calculations are also included. You should work through this material in the way that best suits you. You may choose to read it aloud to yourself or a friend, or to write it out in longhand. The material in this book gives you the essential points of the topic, but remember that to revise effectively you will need to rework it, either mentally or on paper, into concise factual notes that you will be able to remember under exam conditions.

Throughout the book you will find tip boxes that will help you to achieve the A* grade. Some of these apply to all topics; others concentrate on a particularly tricky piece of theory, or something that often causes candidates to trip up.

Towards the end of each section is a review box that acts as a checklist. Once you have worked through the section, check that you can do everything listed. If not, use the page references to refer back to the text and revise that part again.

Each section concludes with a set of practice questions. These are written in exactly the style you will encounter in your exam paper. The answers are provided at the back of the book. Practice makes perfect, so once you have completed the questions in this book, get hold of some actual past papers.

Note: This guide uses the spelling of 'sulfur' with an 'f'. Some exam boards will use the alternative spelling 'sulphur', therefore you should check prior to your examination to ensure you are using the correct spelling.

Key words have been printed in bold: be sure you understand their meaning.

How will I be assessed?

Edexcel

Edexcel candidates take two exam papers:

- Paper 1 lasts two hours and is worth two-thirds of the overall mark.
- Paper 2 lasts one hour and is worth one-third of the overall mark.

Edexcel candidates are not assessed through coursework.

Successful candidates must meet all three assessment objectives:

AO1 Knowledge and understanding

AO2 Application of knowledge and understanding, analysis and evaluation

AO3 Investigative skills (from June 2013 AO3 will change to: Experimental skills, analysis and evaluation of data methods).

CIE

CIE candidates take two papers from the following:

- Paper 1, a core curriculum multiple choice paper lasting 45 minutes and worth 30% of the overall mark.
  and either
- Paper 2, a core curriculum paper lasting one hour 15 minutes and worth 50% of the overall mark.
  or
  Paper 3, an extended curriculum paper lasting one hour 15 minutes and worth 50% of the overall mark.

Plus **one** of the following, all of which are worth 20% of the overall mark:

- Paper 4, coursework conducted at your school.
- Paper 5, a practical test lasting one hour 15 minutes, conducted at your school.
- Paper 6, a written paper on practical theory lasting one hour.

Successful candidates must meet all three assessment objectives:

A Knowledge with understanding

B Handling information and problem solving

C Experimental skills and investigations

Some help with revision

The most common error is to equate success at revision with the time spent on the task. The following advice should help you to revise effectively:

- Never work for more than 30 minutes at a stretch. Take a break.
- Don't revise all day. Divide the day into thirds (you will have to get up at a decent hour) and work two-thirds of the day at most.
- Always start your revision where you finished your last session. You absorb facts better if you always meet them in more or less the same order.
- Don't revise what you already know. That's like practising a complete piano piece when there is only a short part of it that is causing you problems.
- Annotate your revision notes and make use of coloured highlighting to indicate areas of particular difficulty.
- Do some of your revision with fellow students. The knowledge that others find parts of the syllabus tricky relieves anxiety, and other students may well have found an effective way to cope, which they can pass on to you.
- Finally, remember that International GCSE Biology has no hidden pitfalls. If you learn the facts and understand the principles, you will secure the high grade you deserve.

Section One

1 The nature and variety of living organisms

A Characteristics of living organisms

You will be expected to:

★ recall that all living organisms:
- require nutrition
- respire
- excrete their waste
- respond to their surroundings
- move
- control their internal conditions
- reproduce
- grow and develop.

TIP Do not be fooled by how short this section is. Its subject matter will appear in all the other sections.

All living organisms show the following characteristics at some time in their lives.

Nutrition

Nutrition supplies the raw materials or energy for growth and tissue repair.

- Plants, some protoctists and some bacteria **photosynthesise**, changing light energy to chemical energy in sugars built from carbon dioxide and water. Mineral ions are also needed to make other molecules from these sugars, e.g. proteins and DNA.
- Animals, fungi, certain protoctists and certain bacteria obtain nutrition by **digesting** carbohydrates, lipids and proteins from dead or living organisms.

Fig. 1a.01: Animals obtain nutrition by digesting carbohydrates, lipids and proteins from the organisms they eat; some animals, such as this wolf, a top predator, are specially adapted to hunt and kill prey

Respiration

Respiration is the release of energy in cells from the breakdown of food molecules. Aerobic respiration requires oxygen, anaerobic respiration does not.

The term **respiration** is used commonly to mean breathing. Be careful not to use this interpretation in your exam, as the definition of the word in this course is the release of energy inside cells from food molecules. If you want to be really clear about the difference, use the term **cellular respiration**.

Excretion

Organisms **excrete**:

- waste products of metabolic reactions (reactions that happen inside the body), e.g. carbon dioxide from respiration, oxygen from photosynthesis, nitrogenous waste from protein breakdown in animals, e.g. urea in humans
- excess materials absorbed into the body that are not needed, e.g. water, mineral ions.

Response to surroundings

Organisms **sense** changes in the environment (stimuli), and **respond** to them, using nervous or hormonal systems.

Movement

Organisms move by changing shape, position or location.

- Many animals move their whole body from place to place (locomotion).
- Most plants can only move parts of their body, e.g. opening and closing flowers in response to light intensity.

Fig. 1a.02: The kind of locomotion pictured above is called brachiating – swinging from tree to tree using the arms

Control of internal conditions

Organisms control their internal conditions, such as:

- core body temperature in humans, other mammals and birds
- excretion of waste products of metabolism, e.g. carbon dioxide from respiration.

This ensures that conditions inside cells remain relatively constant, so that cell processes work as efficiently as possible. Changes in conditions in cells can affect enzymes and slow down cell metabolic reactions.

Reproduction

Organisms **reproduce**, which means they make more of the same kind of organism.

- Asexual reproduction: new cells are produced by division of a body cell (mitosis). Only one parent is needed.
- Sexual reproduction: male gamete and female gamete fuse to form a zygote. Needs one male and one female parent.

Fig. 1a.03: Polar bears, like all vertebrates, reproduce sexually; the female bear mates with a male before the winter, giving birth to one or two cubs during hibernation

Growth

Organisms grow and develop.

- Multicellular organisms increase in cell number by cell division (mitosis).
- Cells differentiate into different kinds of cells as they develop in different tissues and organs.

TIP

A mnemonic like MRS NERG may help you remember these characteristics:
Movement,
Respiration,
Sensitivity (*response to surroundings and control of internal conditions*),
Nutrition,
Excretion,
Reproduction and
Growth.

You should now be able to:

★ describe and give examples of the main characteristics of living organisms.

Practice questions

1. Living organisms share basic characteristics. Which of these characteristics best describes the following? Some answers require more than one characteristic. Explain each characteristic you have chosen.

   (a) photosynthesis in a cactus **(2)**

   (b) a dog shivering on a cold day **(2)**

   (c) a whale breathing out carbon dioxide **(2)**

   (d) a seed sprouting **(2)**

   (e) a yeast cell dividing **(2)**

TIP

Always look at the number of marks for each question. This shows how many scientific facts or explanations you need to include.

Watch out for questions like this one. Do not start answering it until you have read the entire question, which will show you that it is a mixture of a low demand question (choose characteristics) and a high demand one (explain your choices). Make choices that you can explain clearly.

B Variety of living organisms

You will be expected to:

- ★ describe common features of the six main groups of organisms and give examples of each group:

  plants, animals, fungi, bacteria, protoctists, viruses
- ★ describe what a *pathogen* is and give an example from fungi, bacteria, protoctists and viruses
- CAM ★ define and describe what is meant by a *binomial system*
- ★ list the main features of the vertebrate groups
- ★ describe different classification systems
- ★ list the main features, and give examples, of the following groups, explaining how they are adapted to their environment: flowering plants, arthropods, annelids, nematodes and molluscs
- ★ use simple dichotomous keys to identify organisms.

Classifying organisms

There are five large groups or **kingdoms** of organisms: plants, animals, fungi, protoctists and bacteria.

	Plants	Animals	Fungi	Protoctists	Bacteria (Monera)
Body/cell structure	multicellular (many cells)	multicellular (many cells), nervous coordination, can move from place to place	most multicellular with multinuclear mycelium made of hyphae, some single-celled (e.g. yeast)	single-celled, microscopic, some are like animal cells, others are like plant cells (algae)	single-celled, microscopic, bacterial cell structure: cell wall, cell membrane, cytoplasm and plasmids, chromosome free in cytoplasm, no nucleus
Source of nutrition	chloroplasts, so can photosynthesise	no chloroplasts, so digest organic material	saprotrophic: extracellular digestion and absorption of organic material, lichens – fungi that live associated with algae	some photosynthesise, others ingest food particles (intracellular digestion), or extracellular digestion	some photosynthesise, others extracellular digestion of dead or living organic material
Cell wall structure	cellulose		chitin		various forms

Classifying organisms continued

	Plants	Animals	Fungi	Protoctists	Bacteria (Monera)
Carbohydrate store	starch or sucrose	often as glycogen	may have glycogen		
Examples	flowering plants e.g. grasses (including cereals such as maize), and herbaceous legumes (such as peas or beans)	mammals (e.g. humans), and insects (e.g. housefly and mosquito)	*Mucor* a mould with a typical hyphal structure, yeast (single-celled)	Amoeba (animal-like), *Chlorella* (plant-like)	*Lactobacillus bulgaricus* (used to produce yoghurt from milk), *Pneumococcus* a spherical bacterium

Viruses

Viruses are tiny particles that are not cellular, and are much smaller than bacteria. Different viruses have different shapes and sizes. They are formed from a protein coat surrounding nucleic acid, which is either DNA or RNA.

They invade all other kinds of cells, including bacterial cells. Once inside, they use the cells' biochemistry to make more virus particles. Host cells die and release viruses.

Pathogens

A **pathogen** is any living organism that can cause disease in another. Examples include:

- fungi such as ringworm fungus that infects human skin
- bacteria such as *Pneumococcus* that causes pneumonia in humans
- protoctists such as *Plasmodium* that causes malaria in humans
- viruses such as tobacco mosaic virus that causes discolouring of the leaves of tobacco plants by preventing the formation of chloroplasts, and in humans the influenza virus that causes flu and the HIV virus that causes AIDS.

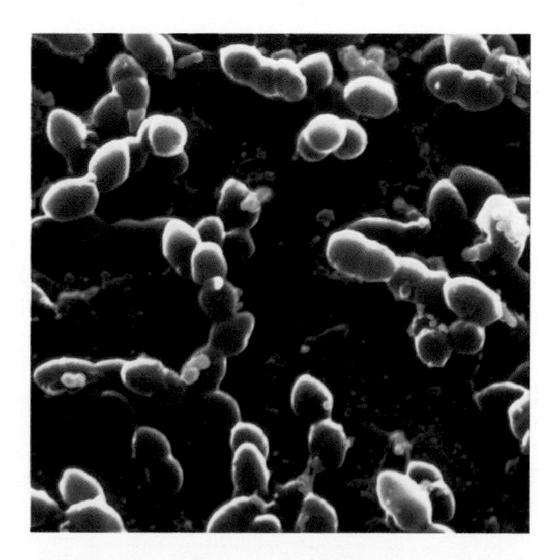

Fig. 1b.01: *Pneumococcus* bacteria: division is by binary fission; mitosis only occurs in cells with a nucleus

CAM

Binomial system

The binomial system gives each species a unique set of genus and species names for accurate identification.

- The names are written in *italics* or are underlined.
- The genus name starts with an upper case letter.
- The species name is always all lower case.

Examples:

- fruit fly, *Drosophila melanogaster*
- yeast, *Saccharomyces cerevisiae*
- maize, *Zea mays*
- human, *Homo sapiens*

The binomial system is international so everyone uses the same name to identify a species.

Vertebrates

Vertebrates have:

- a cranium that protects the brain
- a segmented, flexible vertebral column made up of vertebrae that protects the spinal cord.

All vertebrates have bone except cartilaginous fish (e.g. sharks, rays) that just have cartilage.

Vertebrates also have characteristic internal organs: kidneys, liver, hormone-secreting (endocrine) organs, heart pumping blood through a closed circulatory system.

The vertebrates include the following groups:

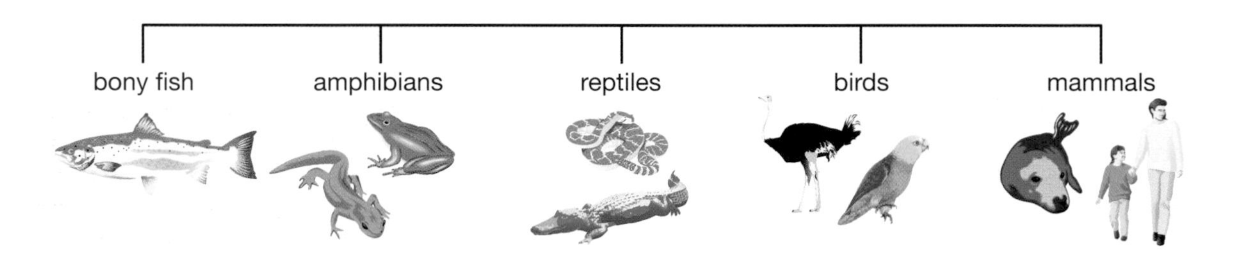

Fig. 1b.02: The main vertebrate groups

Each group in the vertebrates has characteristic features (pages 9-10).

TIP Not all species within a classificatory group will have all the characteristic features. Some will have adapted features suited to the environment in which they live. Details of skeleton structure etc. help classify these species into the correct group.

Bony fish

Bony fish live in water. Their characteristics include:

- fins and tail for swimming
- gas-filled swim bladders – by regulating gases in the swim bladder, they control buoyancy
- bony operculum over gills that allows them to pump water over their gills
- external fertilisation (eggs fertilised in water by sperm)
- body temperature is that of their surroundings (ectotherms).

There are two groups of bony fish:

- ray-finned fish (fins are bony) e.g. barracuda, tuna, pike and goldfish
- lobe-finned fish (fins contain muscle and bone) e.g. coelocanth and lungfish.

Amphibians

Many amphibians are partly adapted to life on land, with a life cycle in two stages:

- aquatic eggs and larvae
- terrestrial adults.

Amphibian characteristics include:

- a thin, moist skin through which they exchange gases with the environment
- external fertilisation
- body temperature is that of their surroundings (ectotherms).

There are three groups of amphibians:

- salamanders and newts (have tails)
- frogs and toads (without tails)
- caecilians (limbless, 'blind' underground amphibians).

Fig. 1b.03: The shiny, wet skin of the red-eyed tree frog is typical of amphibians; their skin is also well supplied with blood vessels to improve the rate of gas exchange and they secrete antimicrobial chemicals to protect their skin from infection

Reptiles

Reptiles generally live on dry land. Their characteristics include:

- scales that protect the surface of the skin
- internal fertilisation (sperm fertilises the egg inside the female's body)
- body temperature is that of their surroundings (ectotherms)
- eggs laid with leathery shells, so they can survive on dry land.

Reptiles include turtles, lizards, snakes and crocodiles.

Scientists are still debating the relationships of the different reptile groups, particularly when birds and fossil reptiles are also considered. For example, crocodiles seem more closely related to birds than to turtles. This part of vertebrate classification may change in the future.

Fig. 1b.04: Python hatching from an egg

Birds

The characteristics of birds include:

- body temperature regulated using heat produced from respiration (endotherms)
- feathers: for flight and down for insulation
- no teeth: beaked mouth
- keeled breastbone for attachment of flight muscles
- lightweight bones, mainly hollow with supporting struts for strength
- internal fertilisation
- lay hard-shelled eggs.

Fig. 1b.05: Birds lay hard-shelled eggs that can withstand the pressure of an incubating parent

Mammals

The characteristics of mammals include:

- body temperature regulated using heat produced from respiration (endotherms)
- mammary glands – the mother produces milk to feed the young
- hair: for insulation
- two sets of teeth: milk teeth and adult teeth
- internal fertilisation
- most mammals give birth to live young (not eggs).

Fig. 1b.06: Female mammals produce milk in their glands for their young

Cladistics

The classification system described above groups organisms into successively smaller groups from kingdom down to genus, and finally to species using key characteristics for each group. **Cladistics** is another system that compares all the characteristics of an organism with those of other organisms to identify how closely related they are. The results are used to produce a diagram like that shown in Fig. 1b.07.

The cladistic diagram in Fig. 1b.07 shows that:

- turtles, snakes, crocodiles, dinosaurs and birds share more characteristics with each other than with mammals
- birds and dinosaurs share the most characteristics.

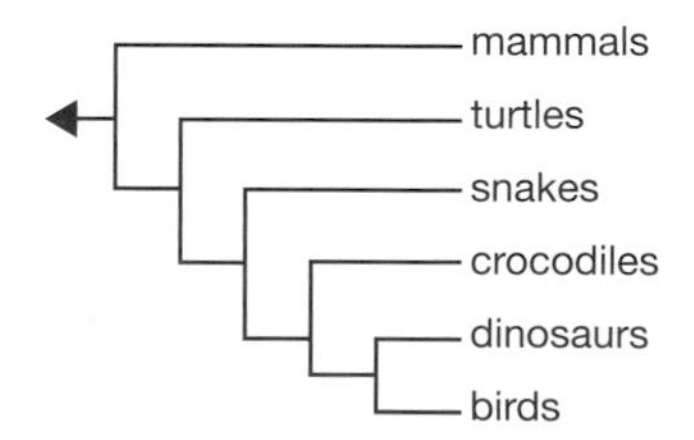

Fig. 1b.07: Cladistic diagram

Cladistic diagrams can be produced using:

- physical characteristics
- DNA (or RNA for bacteria) sequences.

The two approaches don't always produce the same results, as in the discussion about reptiles on page 9.

Flowering plants

The main feature of flowering plants is the flower, which contains all the reproductive structures of the plant.

The two main groups of flowering plants are monocotyledons and eudicotyledons (sometimes just called dicotyledons because only a few species are not eudicotyledons).

Fig. 1b.08: Flowering plant

Monocotyledons

Monocotyledons have:

- one embryonic seed leaf (cotyledon)
- parallel veins in their leaves
- petals in multiples of three
- vascular bundles (contain xylem and phloem) scattered throughout the stem
- vascular tissue in root that is a cylinder around a non-conducting central core
- roots that tend to be fine and fibrous.

Examples of monocotyledons:

- grasses (including cereals)
- palms
- orchids
- bulb plants such as onions, garlic, daffodils, tulips and lilies.

Fig. 1b.09: Maize and daffodil, both examples of monocotyledons

Eudicotyledons

Eudicotyledons have:

- two embryonic seed leaves (cotyledons)
- branching veins in their leaves
- petals usually in multiples of four or five
- vascular bundles arranged in a ring underneath the stem's surface
- vascular tissue in root that forms a star-like arrangement
- root systems usually made up of one or more large roots.

Examples of eudicotyledons:

- roses
- beans
- cabbages
- broad-leaved trees such as oaks
- daisies.

Fig. 1b.10: A rose and an oak tree, both examples of dicotyledons

Invertebrates

All animals that are not vertebrates are classified as invertebrates. These animals include the arthropods, annelids, nematodes and molluscs.

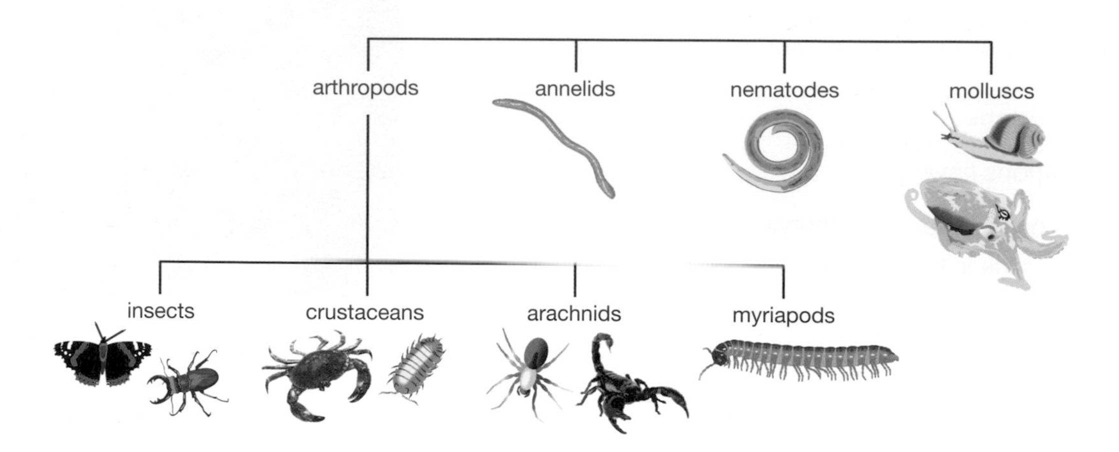

Fig. 1b.11: The main invertebrate groups

Arthropods

The characteristics of arthropods include:

- exoskeleton (hardened shell outside the body) made up of layers of chitin and protein, secreted by cells underneath the exoskeleton
- jointed legs
- segmented bodies (so they can move)
- open circulation system – tissues bathed in 'blood', no blood vessels.

The arthropods include four major sub-groups.

- **Insects** usually have 3 pairs of legs, 2 pairs of wings and a body made up of 3 parts. They include butterflies, bees and beetles. Many can fly and many are important pollinators of plants. The mouth parts of the housefly and mosquito are adapted to their particular needs: the housefly coats its food with saliva and sucks up through sponge-like mouthparts; the mosquito pierces the skin and sucks up the protein-rich blood it needs.
- **Crustaceans** have 2 pairs of antennae, hardened mouthparts for biting and grinding, and legs made of 2 parts. They include lobsters, crabs, woodlice, barnacles and brine shrimps. Most are aquatic and the larval stages disperse in water currents.
- **Arachnids** have no antennae, 4 pairs of legs and no jaws and so cannot chew. They include spiders, mites and scorpions. Many are predators that subdue their prey with poison before injecting them with enzymes to dissolve the tissue.
- **Myriapods** have a body with many segments. Millipedes have 2 pairs of legs per segment. Centipedes have 1 pair of legs per segment. Millipedes are herbivores. Centipedes are active predators.

Annelids

The characteristics of annelids include:

- long bodies of identical segments (metameres)
- specialised excretory organs (nephridia)
- closed circulatory system – 'blood' contained in blood vessels
- skin as the gas exchange surface.

They include earthworms which are important because they help break down dead organic material in the soil.

Fig. 1b.12: An earthworm's segments are clear, and appear pink due to the haemoglobin dissolved in its blood

Nematodes

Nematodes are unsegmented, worm-shaped animals. They are often called roundworms. They have:

- thin skin covered by a protective layer (cuticle), and bristles for sensing touch
- a mouth that may have small teeth or a sharp point for piercing tissue
- no specialised excretory system
- a nervous system of four nerves.

Over half the species known are free-living, but many are parasitic, e.g. hookworms, whip worms, thread worms, filarial worms, *Trichanella spiralis* (in uncooked pork, causes trichinosis in humans).

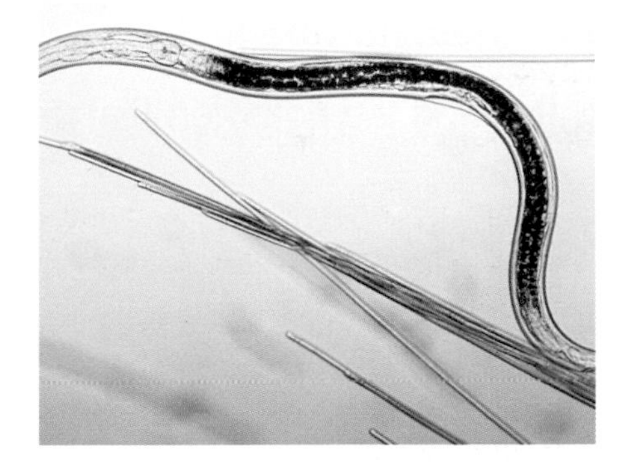

Fig. 1b.13: A free-living aquatic nematode pictured here with algae

Molluscs

Molluscs show a great variety in body structure but the key characteristics are:

- a main body (visceral mass) with single 'foot' attached
- mantle on top of the visceral mass that secretes a shell and contains the respiratory structures, either gills or lung
- a mouth and anus that usually both open into mantle cavity
- specialised excretory structures (nephridia)
- an open circulation system – no blood vessels
- rasping tongue (radula) with spines made of chitin.

Examples of molluscs include:

- gastropods such as snails and slugs that live on land and in water, and rasp plant material with their radula, a tongue covered in tiny spines
- bivalves (two-shelled) such as clams that filter food out of the water where they live
- cephalopods – soft-bodied active predators such as squid and octopus.

Fig. 1b.14: The snail moves up plant stems along a layer of secreted mucus

Identification keys

We can use easily identifiable features of an organism to produce an identification key. When a key is presented in the form of two choices (e.g. yes or no) at each stage, it is called a **dichotomous key**.

A dichotomous key for animals might start like this:

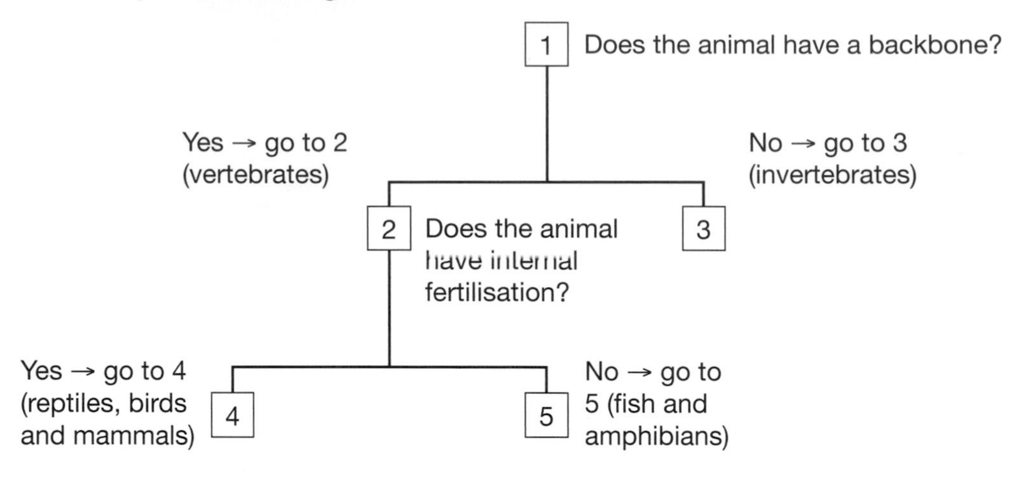

You should now be able to:

★ describe the common features of each of the following groups:
- plants
- animals
- fungi
- bacteria
- protoctists
- viruses (see page 2)

★ define the term *pathogen* and give an example from (a) fungi, (b) bacteria, (c) protoctists and (d) viruses (see page 7)

CAM ★ define the *binomial system* for naming organisms (see page 8)

★ list the main features of the following vertebrate groups:
- bony fish
- amphibians
- reptiles
- birds
- mammals (see pages 8–10)

★ list the main features used for classifying the following:
- flowering plants, including monocotyledons and eudicotyledons (see page 11)
- arthropods, including insects, crustaceans, arachnids and myriapods (see page 12)
- annelids (see page 13)
- nematodes (see page 13)
- molluscs (see page 13)

★ describe the use of simple dichotomous keys for identifying organisms (see page 14).

Practice questions

1. Explain why a tree is in the plant kingdom and a bird is in the animal kingdom. **(8)**
2. List the main differences between vertebrates and invertebrates. **(6)**
3. What two levels of classification are used to name an organism in the binomial system? **(2)**
4. Which three kingdoms have the substance chitin somewhere in their bodies? **(3)**

Section Two

2 Structures and functions in living organisms

A Levels of organisation

You will be expected to:

★ describe the levels of organisation within organisms: organelles > cells > tissues > organs > organ systems

CAM ★ relate the structure of the following cells to their functions:
- ciliated cells
- root hair cells
- xylem vessels
- muscle cells
- red blood cells.

Organisms have different levels of organisation within their bodies, which helps them to carry out the life processes as effectively as possible.

Organelles

Organelles are membrane-bound structures within cells that are separated from the cytoplasm, and carry out basic functions. They include the nucleus, plant vacuole and chloroplasts. These are described in more detail in Section 2B Cell structure (see page 21).

Cells

Cells are the smallest unit of an organism able to carry out all of the functions of life.

In multicellular organisms, most cells perform some special functions that other cells do not. Examples are described in Section 2B Cell structure (see page 21).

Cell function is regulated by homeostasis (see page 93), and can be affected by molecules from other parts of the organism (e.g. hormones) or other organisms (e.g. pheromones).

Specialist cells

Cells	Where found	Function	Adaptation
ciliated cells	human respiratory tract *Paramecium*	to move substances or to locomote	tiny hairs move together to move a liquid or move through a liquid
root hair cells	near ends of roots in plants	absorption of water and dissolved solutes from soil water	long thin shape gives large surface area
xylem vessels	vascular bundles of plants	conduction of water through plant and support of plant	form long tubes through plant, thick walls are strong
muscle cells	muscle tissue	contraction to move skeleton, blood or other materials	able to shorten in length
red blood cells	blood	transport of oxygen	contain haemoglobin that binds to oxygen

Tissues

A **tissue** consists of groups of similar cells that work together to perform a shared function. Tissues only occur in multicellular organisms.

Plants have three main types of tissue:

- dermal tissue that forms a protective outer covering, including guard cells, root hair cells, epidermal cells
- vascular tissue that provides mechanical support to a plant and transports water and solutes through the plant, including phloem cells and xylem cells
- ground tissue, which is the packing and supportive tissue around other tissues, and which also stores substances such as starch (e.g. in a potato)
- meristem tissue that divides by mitosis.

Animals have four main types of tissue:

- muscular tissue, which can contract and generate forces, e.g. skeletal muscle that moves the skeleton, cardiac muscle that pumps blood out of the heart, and smooth muscle that controls movement through tubes such as the digestive tract
- nervous tissue, which initiates and transmits nerve impulses in neurones in nerves, the brain and spinal cord
- epithelial tissue, which lines external and internal surfaces and may include cells specialised in the exchange of materials, such as in the lungs or small intestine, or in secretion, as found in the pancreas, testes and ovaries
- connective tissue that connects, anchors and supports tissues and organs.

Stem cells are found throughout animal tissues.

Organs

An **organ** is made up of at least two different kinds of tissue that perform specific and related functions. Examples of organs include:

- in animals: heart, lungs, liver, brain, kidney
- in plants: leaf, stem, root.

Organ systems

Organs that work together to perform particular body functions are called organ systems, for example:

- the brain, spinal cord and sensory organs are parts of the nervous system
- the heart and blood vessels are parts of the circulatory system
- the kidneys, ureters, urethra and bladder are organs of the excretory system
- the mouth, oesophagus, stomach, pancreas, and small and large intestines are organs of the digestive system.

The output of one part of a system (which can include material, energy or information) can become the input to other parts. Systems work together to carry out all the life processes in living organisms.

You should now be able to:

- ★ describe the levels of organisation within organisms (see page 18)
- CAM ★ explain how the structure of the following cells is related to their function: ciliated cells, root hair cells, xylem vessels, muscle cells, red blood cells (see page 19).

Practice questions

1. Using an organ system of your choice, give examples of the organs, tissues and cells in that system and explain their contribution to the functioning of the system. **(8)**

B Cell structure

You will be expected to:

- ★ recognise cell structures, such as the nucleus, cytoplasm, cell membrane, cell wall, chloroplast and vacuole
- ★ describe the functions of the cell structures listed above
- ★ describe differences between plant and animal cells
- CAM ★ calculate the magnification and size of biological specimens.

The structures in cells are specialised to carry out different basic functions.

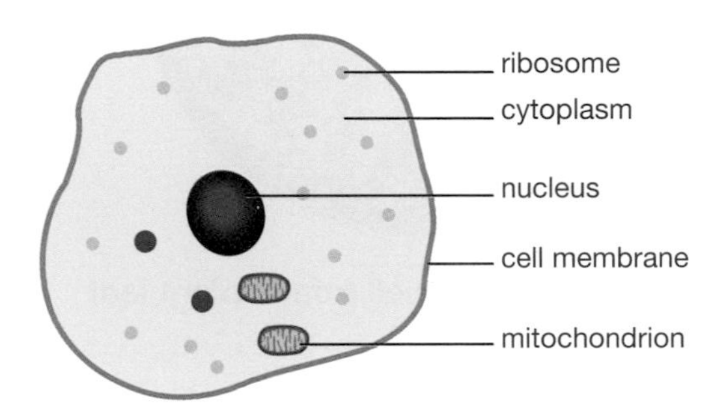

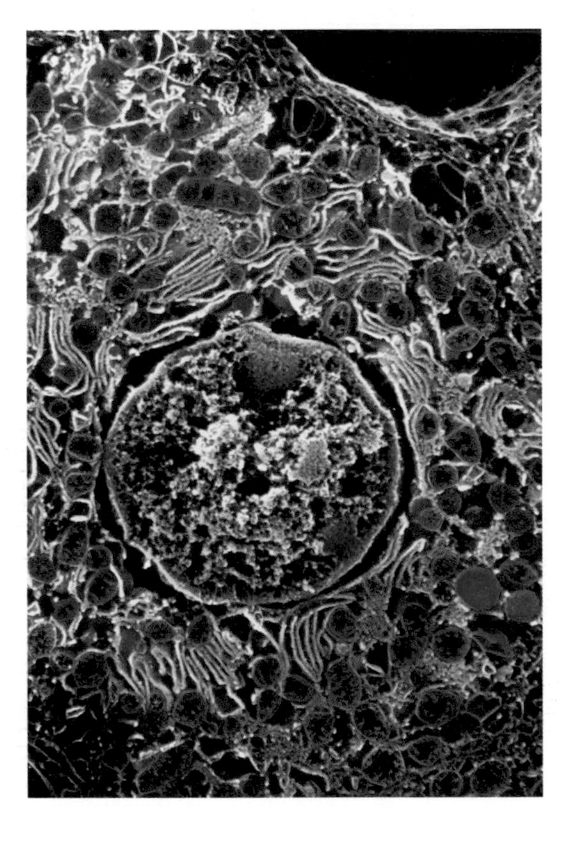

Fig. 2b.01: (Left) Diagram of a generalised animal cell (Right) A micrograph of a liver cell

Animal cells have the following structures that are visible under a light microscope:

- **nucleus** – surrounded by a membrane to separate it from the cytoplasm, contains the genetic information
- **cell membrane** – surrounds the whole cell, controls what can enter and leave the cell
- **cytoplasm** – a thick liquid inside the cell membrane in which other cell structures are held, also contains dissolved solutes
- **mitochondrion** – site of aerobic respiration and therefore where most energy is made.

Animal cells also contain:

- **ribosome –** site of protein synthesis i.e. where enzymes, antibodies and other proteins are made.

Plant cells have the same structures as animal cells. In addition, the following structures can also be seen under a light microscope in plant cells:

- **chloroplasts** – structures surrounded by a membrane that contain chlorophyll, where photosynthesis takes place
- **cell wall** – surrounds the cell membrane, made of cellulose, gives the cell shape and prevents bursting
- large **central vacuole** – contains dissolved solutes some of which are stored temporarily, such as sucrose.

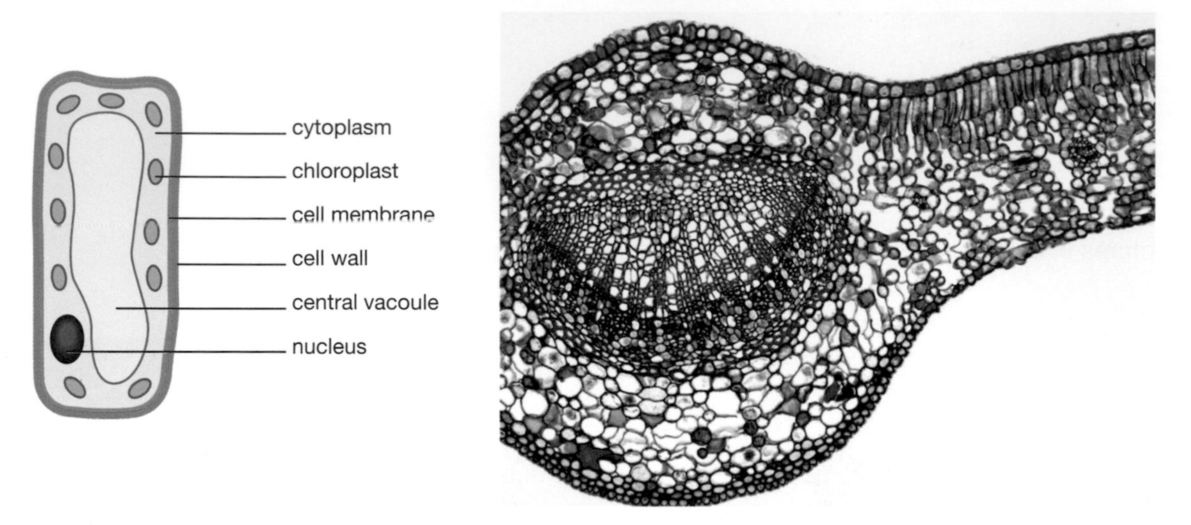

Fig. 2b.02: The cell structures of a palisade cell from a plant leaf

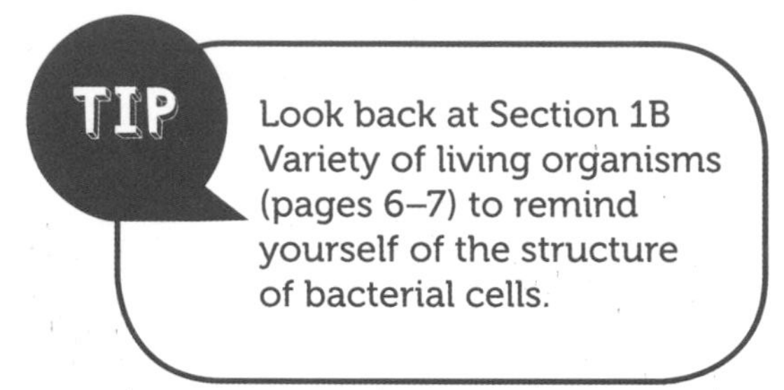

CAM

Calculating magnification and size

The **magnification** of a specimen that is being viewed under a microscope is calculated by multiplying together the magnification of the eyepiece and the objective being used. For example:

- if the eyepiece is ×10 and the objective is ×20
- the magnification is 10 × 20 = 200.

The actual size of a structure on a slide can be calculated using the formula:

$$\text{actual size} = \frac{\text{measured size}}{\text{magnification}}$$

The measured size can be found using a slide **graticule**.

For example:

- if the measured size is 4 mm and the magnification is 200
- the actual size is $\frac{4}{200}$ = **0.02 mm**

You should now be able to:

- ★ describe the structure and function of the following structures in cells: nucleus, cytoplasm, cell membrane, cell wall, chloroplast and vacuole (see pages 21–22)
- ★ describe the differences between a typical animal cell and a generalised plant cell (see pages 21–22)
- CAM ★ calculate the magnification and size of biological specimens (see page 22).

Practice questions

1. There are 20 onion cells in a drawing measuring 5 cm across. The magnification is x100.

(a) Calculate the actual size of the 20 onion cells in mm. **(2)**

(b) Onion cells are plant cells. What three structures do they have that animal cells do not have? **(2)**

(c) When left in the light, onion cells turn green. What is this substance and what does it do? **(2)**

(d) Describe two differences between an onion skin cell and a root hair cell. **(2)**

2. Use the table provided to list all of the structures that are common to both animal and plant cells, *and* the function of each structure.

	Cell structure	Cell structure's function	
1			**(1)**
2			**(1)**
3			**(1)**

C Biological molecules

You will be expected to:

★ recall the elements present in carbohydrates, proteins and lipids
★ name the sub-units that make up carbohydrates, proteins and lipids
★ describe the tests for glucose and for starch
CAM ★ describe the tests for protein and for fats
★ explain how enzymes act as biological catalysts in metabolic reactions
★ describe how enzyme function can be affected by temperature and pH
★ describe simple experiments to show how enzyme activity is affected by temperature
CAM ★ explain enzyme action using the lock and key model, including the effect of temperature and pH on enzyme activity
★ describe the use of enzymes in biological washing products and in the food industry.

Elements in biological molecules

- **Carbohydrates** contain only carbon, hydrogen and oxygen in the following proportions: $C_n(H_2O)_n$ (where *n* can be any number). For example, the soluble, simple sugar glucose is $C_6(H_2O)_6$. Other simple sugars are galactose and fructose. Complex carbohydrates include insoluble starch, glycogen and cellulose.
- **Proteins** contain carbon, hydrogen, oxygen, as well as nitrogen. Some proteins also contain sulfur.
- **Lipids** are fats (solid lipids) and oils (liquid lipids); they contain carbon, hydrogen and oxygen, but they have a much higher proportion of hydrogen than oxygen.

Structures of biological molecules

Carbohydrates

The complex carbohydrates you will come across in your course are starch and glycogen. They are large insoluble molecules, made up of many glucose molecules joined together.

- Starch is mostly a straight chain of glucose molecules.
- Glycogen is a branching chain of glucose molecules.

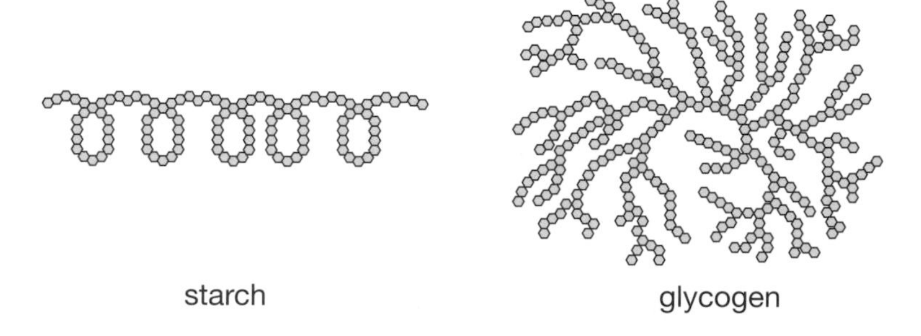

Fig. 2c.01: Structure of part of a starch molecule and a glycogen molecule; each small hexagon in the molecules is a glucose molecule

There are also smaller carbohydrates:

- maltose is two glucose molecules joined together
- lactose is one glucose molecule and another simple sugar molecule (galactose) joined together
- sucrose (table sugar) is one glucose molecule and another simple sugar molecule (fructose) joined together.

Proteins

Proteins are made up of one or more chains of amino acids joined together.

There are 20 different amino acids, with the same basic structure. The order of amino acids varies in different proteins, so the 20 amino acids can make thousands of different proteins.

Each type of protein is folded into a specific 3D (three-dimensional) shape depending on how the amino acids and environment affect each other. The function of each protein depends on its shape.

Examples of proteins covered in your course include:

- proteins in cell membranes that control movement of substances during active transport
- enzymes that control cell reactions, such as in respiration, photosynthesis, DNA copying and digestion
- hormones that control cell responses, such as insulin, oestrogen and testosterone
- antibodies in the blood that help control infection.

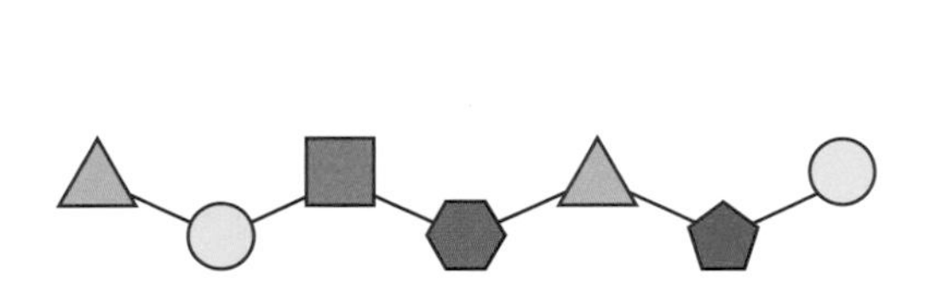

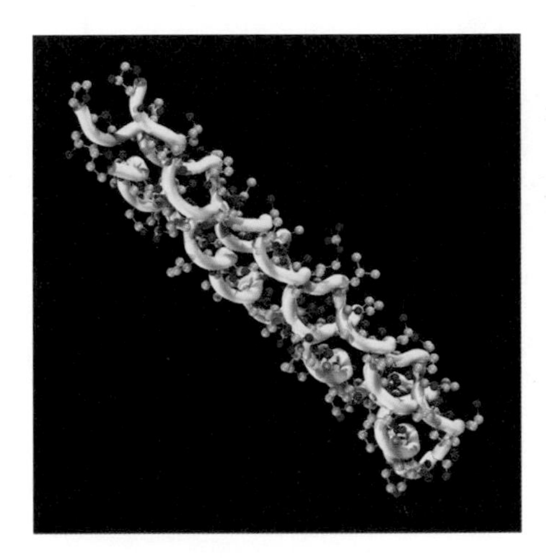

Fig. 2c.02: (Left) Part of a protein molecule; each different shape represents a different amino acid, and is not the real shape of amino acids (Right) A computerised model of a protein molecule

Lipids

Lipids are a large and diverse group of chemicals that include steroid hormones, cholesterol, vitamin A and vitamin D, fats and oils. A single fat or oil molecule is made up of a glycerol molecule to which three fatty acid molecules are attached.

Different lipids contain different fatty acid molecules.

Fig. 2c.03: Basic structure of a lipid molecule; the blue shape is a glycerol molecule, and the red 'tails' are fatty acid molecules

Tests for biological molecules

Test for glucose

Benedict's solution is blue. Some sugars, like glucose, fructose, lactose and maltose, are called **reducing sugars** because they reduce the chemicals in Benedict's solution.

- Add an equal amount of the test solution and Benedict's solution to a test tube.
- Shake to mix and then bring gently to the boil. (Shake or stir continuously to minimise spitting.)
- After 2–3 minutes, if the test solution contains a simple sugar (e.g. glucose), the blue colour changes. The colour change depends on the concentration of the sugar:

(low glucose concentration) green > yellow > brick red (high glucose concentration)

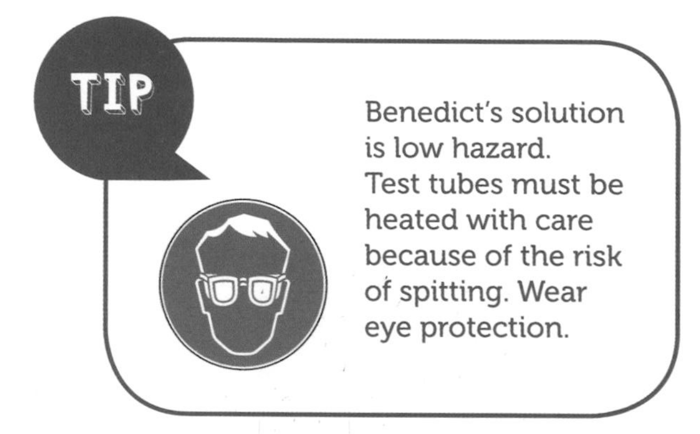

The copper(II) sulfate in Benedict's solution is what makes it blue. Simple sugars like glucose, heated with the solution, reduce the copper(II) sulfate to copper(I) oxide, changing the colour.

Sucrose will *not* give a positive result because it is a non-reducing sugar. However, it will react if it is first broken down to glucose and fructose by boiling it with hydrochloric acid.

Fig. 2c.04: Test for glucose: negative on the left, positive on the right

Test for starch

Iodine solution is a clear dark orange. When mixed with a solution containing starch, a blue–black precipitate is formed. This test is often used with leaves to show whether photosynthesis has been taking place.

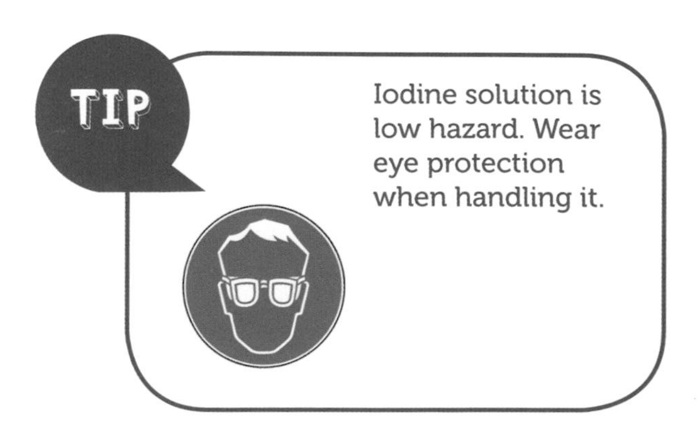

Fig. 2c.05: Test for starch: negative on the left, positive on the right

Test for protein

One test for protein is called the **biuret test**.

- Add equal volumes of the test solution and 5% sodium hydroxide solution to a test tube and mix.
- Add 2 drops of 1% copper sulfate solution and mix.
- If the test solution contains protein, a mauve or purple colour will slowly develop in the solution.

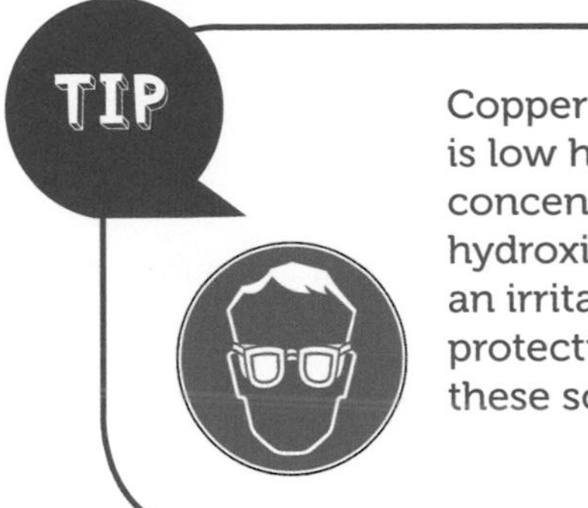

Copper sulfate solution is low hazard at this concentration, sodium hydroxide solution is an irritant. Wear eye protection when using these solutions.

Test for fats

Lipids dissolve easily in ethanol, but not in water.

- Add equal amounts of the test solution and absolute ethanol to a test tube.
- Cover the end of the tube and shake it vigorously.
- Add an equal volume of cold water.
- If lipid is present in the test solution, the mixture will turn cloudy.

Ethanol is flammable. Keep away from open flames, such as a Bunsen burner.

Enzymes

Enzymes are proteins that act as **biological catalysts**. This means they increase the rate of metabolic reactions in the body without being changed when the reaction has finished.

Enzymes make it easier for reactions to happen (by lowering the activation energy of the reaction). This means the reactions can occur fast enough at normal body temperatures to sustain life processes.

Enzymes are effective in small concentrations and are reusable. Enzymes are specific. Each enzyme only catalyses one kind of reaction because only a certain substrate will fit into an enzyme's active site.

Temperature, pH and enzymes

Enzymes are affected by temperature.

- An enzyme has an **optimum temperature** at which the reaction it catalyses proceeds at its fastest rate.
- The rate of an enzyme-controlled reaction is slower when the temperature is lower than this optimum.
- Above the optimum temperature, the enzyme starts to change shape and becomes **denatured**, so that it cannot work as well.

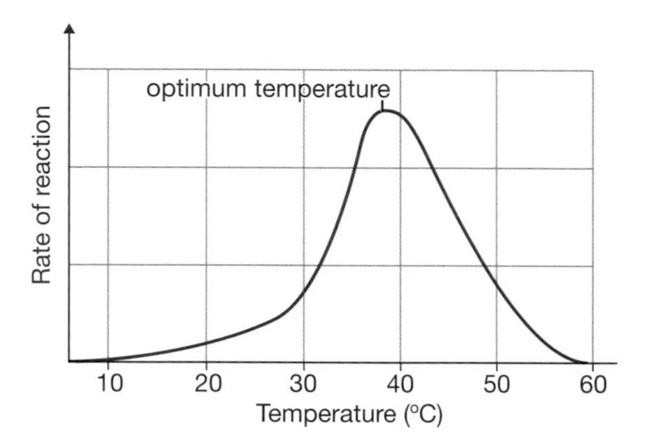

Fig. 2c.06: The effect of temperature on an enzyme-controlled reaction

Enzymes that work in the core of the human body, such as in the digestive system, usually have an optimum temperature of around 37 °C.

Enzymes may also be affected by pH, having an optimum pH above and below which the reaction proceeds more slowly. Different enzymes may have a different optimum pH depending on where in the body they work.

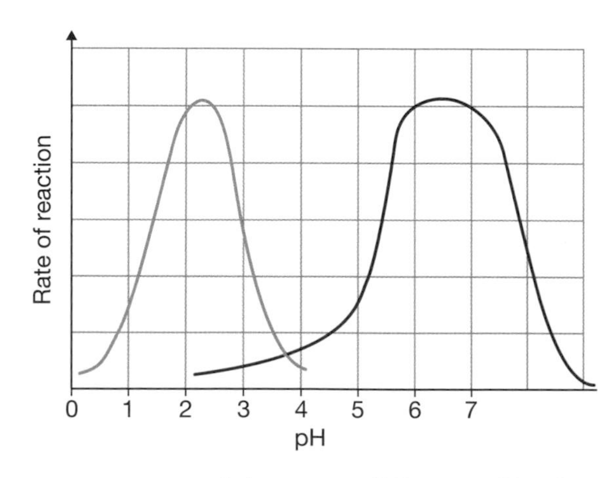

Fig. 2c.07: Different enzymes work better at different pHs; the green line shows an enzyme that works in the acidic conditions of the stomach; the red line shows an enzyme that works in the more alkaline conditions of the small intestine

Simple experiment to show the effect of temperature on enzyme activity

To show how enzyme activity can be affected by changes in temperature, we can use the breakdown of starch to maltose by the enzyme amylase.

- Prepare a spotting tile with one drop of iodine solution in each dimple.
- Label four test tubes *control*.
- Label four other test tubes *starch + amylase*. Measure 10 cm^3 of 2% starch solution into each of the eight test tubes. You will add the amylase later.
- Label four other tubes *amylase* and put 1 cm^3 of 1% amylase solution in each tube.
- Place one control tube, one starch + amylase tube, and one amylase tube into water baths set at different temperatures, e.g. ice water bath, room temperature tap water, 40 °C hot water bath, 60 °C hot water bath. (Use thermometers in the water baths to record temperatures accurately. Avoid the risk of scalding with the hot water baths.)
- Leave the tubes for 5 minutes so that the contents of each control tube reaches the temperature of the water baths it is in.
- To each tube labelled starch + amylase add 1 cm^3 of 1% amylase solution from the test tubes labelled *amylase*.
- To each control tube add 1 cm^3 of water from the water bath.
- Every 30 seconds remove a drop from the control and starch + amylase test tubes, and add to a drop of iodine solution on the spotting tile.
- Compare the reaction of the control and starch + amylase tubes at each temperature to identify when the contents of the starch + amylase tube stops reacting with the iodine solution – this is the point when the amylase has digested all the starch.

The starch + amylase tube at the temperature closest to the optimum for amylase will be the first to stop forming the blue–black precipitate with the iodine.

Simple experiment for an enzyme and pH

The experiment described, for temperature on the previous page, can be adjusted to find the effect of pH on amylase.

- Prepare a spotting tile with one drop of iodine solution in each dimple.
- Have prepared five test tubes containing buffer solutions at different pHs, one tube for each pH (e.g. ranging from pH2–10). Label the test tubes with their pH values. Add 1cm^3 of the amylase solution to each test tube with 1cm^3 of 1% amylase.
- Measure 10 cm^3 of 2% starch solution into five new test tubes labelled *starch*.
- Place all tubes in a rack in a warm water bath (c. 37 °C), and leave for 5 minutes so the contents reach the temperature of the water bath.
- Add 1 cm^3 of 1% amylase solution to each starch tube and mix the contents.
- Every 30 seconds remove a drop from *each* of the 5 starch + amylase test tubes and add it to a drop of iodine solution on the spotting tile.

The tube at the pH closest to the optimum for amylase will be the first to stop forming the blue–black precipitate with the iodine.

Lock and key model of enzyme action

The **lock and key model** helps to explain how enzymes work. The name comes from the idea that the enzyme has an **active site**, with a shape that matches the substrate molecule (or molecules).

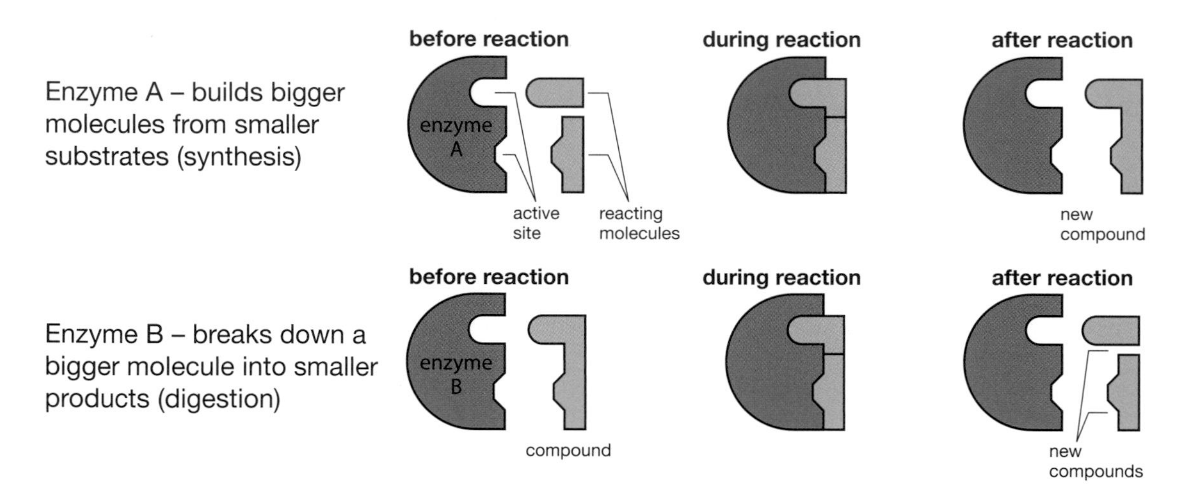

Fig. 2c.08: The lock and key model

The substrate molecule fits into the active site, forming an enzyme–substrate complex. The reaction is catalysed, forming the products, and the products are released.

Explaining the effect of temperature on enzymes

The active site of an enzyme is a 3D shape. The reaction works best when the substrate fits it properly.

- Increasing temperature above the optimum starts to change the shape of the active site. The substrate no longer fits well, so the rate of reaction slows.
- Above a particular temperature, the shape has changed so much that the enzyme–substrate complex no longer forms. This is when the enzyme is denatured. The reaction is no longer catalysed, and no more products are formed.
- At low temperatures the movement (kinetic) energy of the molecules is low. Therefore fewer enzyme molecules and substrate molecules collide within the same time. So, fewer reactions happen in the same time.

Explaining the effect of pH on enzymes

The pH of the solutions surrounding an enzyme can alter the 3D shape of the active site. Changing the shape makes it less easy for the enzyme–substrate complex to form, so the rate of reaction slows down.

Uses of enzymes

We use enzymes for many purposes.

- Biological washing detergents. Proteases, lipases and amylases may be added to washing detergents to help break down biological compounds that form stains, such as blood, body oils and grass stains, at lower temperatures and more quickly than without enzymes. This saves time and energy, and reduces wear on clothes during washing.
- When fruit (e.g. apple) is pressed to extract the juice, plant cells are also filtered into the juice. This makes the juice cloudy and keeps some of the juice trapped within the plant cells. **Pectinase** is an enzyme that breaks down some of the chemicals in plant cell walls. If the apple pulp is treated with pectinase before squeezing, more juice is produced from the same amount of pulp, and the juice is clear.

Fig. 2c.09: When apple pulp is treated with pectinase, the juice is clear

You should now be able to:

- ★ name the chemical elements in carbohydrates, proteins and lipids (fats and oils) (see page 24)
- ★ describe the structure of carbohydrates, proteins and lipids in terms of their basic units (see pages 24–25)
- ★ describe tests for glucose and starch (see page 26)
- CAM ★ describe tests for protein and fats (see page 27)
- ★ define the terms *catalyst* and *enzymes* (see page 28)
- ★ describe how enzyme activity can be affected by changes in temperature and pH (see page 28)
- ★ describe a simple experiment to illustrate how enzyme activity is affected by temperature (see page 29)
- CAM ★ explain enzyme action in terms of the lock and key model (see page 30)
- ★ state two uses of enzymes and explain why they are used (see page 31).

Practice questions

1. Put a ✓ in the appropriate boxes.

	Carbohydrates	Proteins	Lipids	
Contains C, H and O				(1)
Contains N				(1)
Sometimes contains S				(1)
This group includes enzymes				(1)
Contains glycerol				(1)
Subunits are amino acids				(1)
Fungi can digest these				(1)
Contains the highest % of high energy C-H bonds				(1)
Made at ribosomes				(1)
Made by photosynthesis				(1)

D Movement of substances into and out of cells

You will be expected to:

★ give simple definitions for the terms *diffusion*, *osmosis* and *active transport*
★ describe the importance of diffusion of gases and solutes, and of water as a solvent
★ describe how substances move in and out of cells by diffusion, osmosis and active transport
★ describe the importance of turgid cells in supporting a plant
CAM ★ describe the effect of osmosis on plant and animal tissues
★ give examples of active transport in plant and animal cells
★ describe the effect of surface area to volume ratio, temperature and concentration gradient on the rate of movement of substances into and out of cells
★ describe simple experiments on diffusion and osmosis in living and non-living systems.

Diffusion

Diffusion is the net movement of molecules from an area of their higher concentration to a region of their lower concentration.

Molecules are in constant, random movement (except at –273 °C, absolute zero).

- Where molecules are not distributed evenly, there are more molecules in a region of higher concentration than in a region of lower concentration.
- So, there will be more molecules moving from the region of higher concentration to the region of lower concentration than in other directions.
- This results in **net movement** from the region of higher concentration to the region of lower concentration. This is **diffusion**. Net movement stops when the overall concentration is equal but the molecules are still moving.

TIP

Molecules are moving randomly in all directions at all times. Where there is a concentration gradient, there will be net movement away from a region of higher concentration. Remember the word 'net' and use it in your definition.

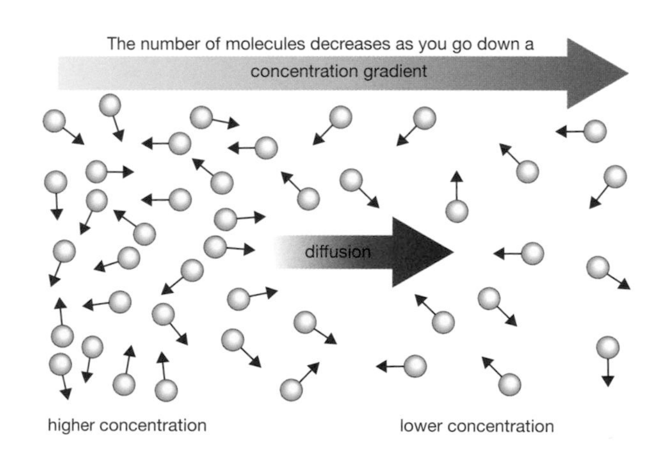

Fig. 2d.01: Diffusion is the net movement of molecules down a concentration gradient

- Diffusion is a passive process – it needs no energy from cell respiration.
- Diffusion is always *down* the **concentration gradient**, from the higher concentration to the lower.

Diffusion can only happen where molecules are free to move, i.e. in gases and in liquids (and in fact in some solids). This makes it possible for:

- gas molecules to diffuse in and out of the air spaces in the spongy mesophyll of leaves via openings in the epidermis of the leaf called stomata. The rate of diffusion can be controlled by guard cells which adjust the size of the stomata. Guard cells are specialised epidermal cells
- solute molecules to diffuse through water, such as in solutions within cells and bodies of organisms, and solutes in soil water or rivers, lakes and oceans. Examples of solutes that diffuse within, and between, cells are oxygen, carbon dioxide, glucose, urea and certain ions.

Diffusion and cells

Cells are surrounded by cell membranes that are **partially permeable**. This means they have tiny holes in them.

Molecules that are small enough can pass through the holes in the cell membrane, e.g. gas molecules such as oxygen and carbon dioxide, water, urea, glucose and ions.

Large molecules, such as sucrose in plants, are too large to diffuse across membranes (which is why they make good transport molecules). Protein molecules are also too large to cross membranes.

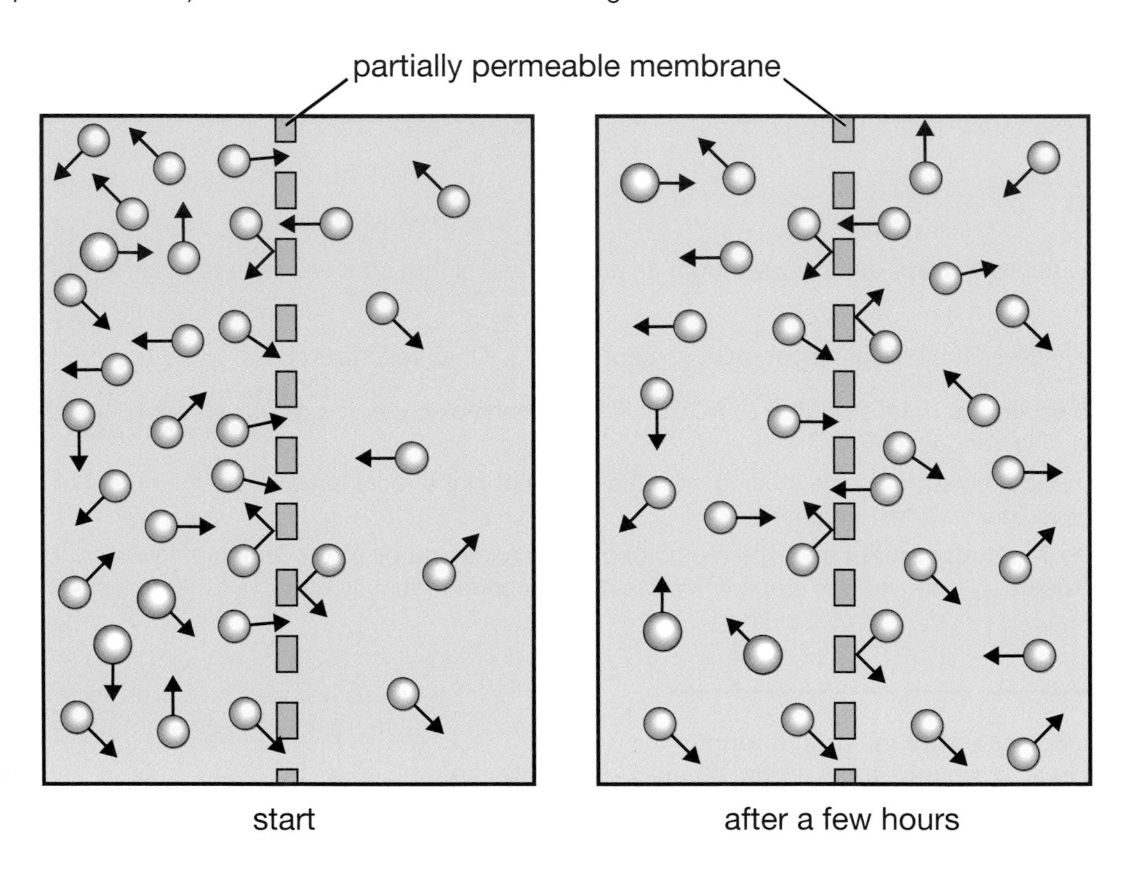

Fig. 2d.02: Diffusion across a partially permeable membrane, such as a cell membrane

Experiments to show diffusion

Non-living systems

- Cubes of agar of different sizes (e.g. sides of 1 cm, 2 cm and 3 cm) are soaked in coloured water for the same length of time.
- When the cubes are cut open, we can see the colour has diffused over a particular distance. For the smallest cube, the colour has diffused as far as the middle.
- This shows that if a cell or organism is too large, diffusion will take too long to move substances to the middle and support all the life processes.

Fig. 2d.03: The cubes were placed in coloured water for the same time and then cut in half

We can also use visking tubing, an artificial partially permeable membrane, to model a cell membrane.

- A solution of starch and glucose is placed inside a tube of visking tubing, and the tube is placed in pure water.
- After about half an hour, the water outside the visking tubing is tested for starch and for glucose.
- The water tests positive for glucose but negative for starch.
- This shows that the glucose molecules are small enough to pass through the partially permeable membrane, but the starch molecules are too large to get through.
- So glucose shows diffusion, but not starch.

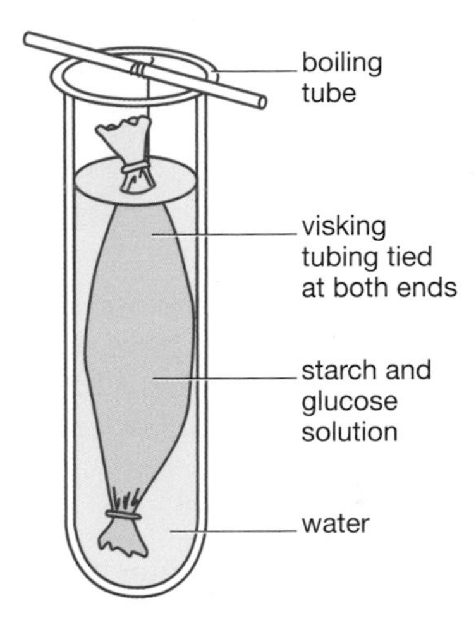

Fig. 2d.04: Apparatus for diffusion experiment

Living systems

A similar experiment to the one above can be done with tissue cut from a potato, or other thick tuber. This shows that diffusion occurs across cell membranes.

Osmosis

Osmosis is a special case of diffusion. It is the net movement of water molecules from a region of their higher water concentration to a region of lower water concentration through a partially permeable membrane. It is also a passive process.

Water molecules are in constant random motion. Where there are more water molecules in a region of higher water concentration than in a region of lower water concentration, there will be *net* movement of water molecules down the water concentration gradient. Net movement stops when the overall concentration is equal, but water molecules are still moving.

TIP

It is easy to get confused when talking about osmosis and concentration. You need to remember to make clear that it is the concentration of **water molecules** that is being considered in osmosis, **not** the concentration of solute molecules as when we talk about diffusion.

Osmosis and animal cells

The rate of osmosis across cell membranes is important for cells.

- If the cell has a lower concentration of water molecules (higher concentration of solute molecules) than the tissues around it, water enters due to osmosis. If too much water enters an animal cell it may burst.
- If the cell has a higher concentration of water molecules (lower concentration of solute molecules) than the tissues around it, water moves out due to osmosis and the cell shrinks. If too much water is lost, the cell may die because water is essential for many cell processes.

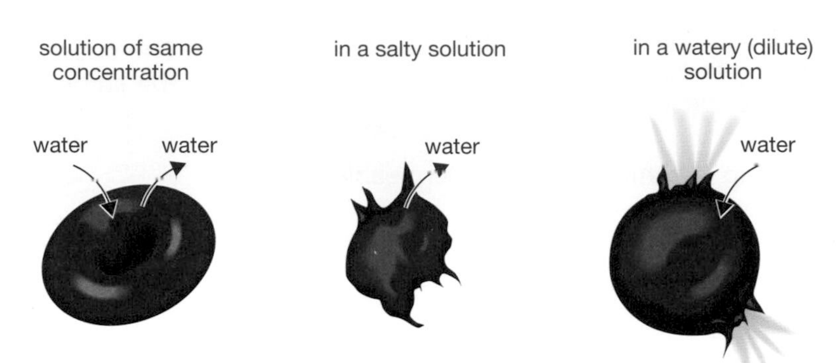

Fig. 2d.05: Red blood cells placed in different solutions (Left) Cell cytoplasm and solution are isotonic and have the same concentration of water molecules (Centre) Cell cytoplasm is hypotonic and has a higher concentration of water molecules than the solution (Right) The solution has higher concentration of water molecules than the hypertonic cell cytoplasm

Osmosis and plant cells

Plant cells gain and lose water by osmosis, as in animal cells. However, a plant cell is surrounded by a tough cellulose wall.

The plant cell wall is completely permeable to water and so does not control the movement of molecules across it, as does the cell membrane.

- The cellulose cell wall stops the cell membrane stretching too far. Therefore, the water pressure inside a turgid cell stops any net movement of water into the cell by osmosis.
- A plant cell that contains as much water as it can is said to be **turgid**.
- The turgid cells of a plant give mechanical support to the plant to keep the stem upright and the leaves held out.
- A plant cell can lose water by osmosis. When the cell membrane pulls away from the cell wall due to the loss of water, the plant cell has plasmolysed. If enough cells in a plant undergo plasmolysis, the plant wilts due to the lack of turgor.

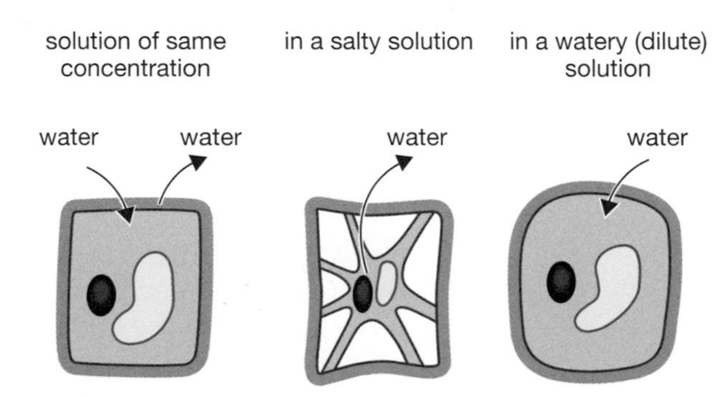

Fig. 2d.06: Plant cells placed in different solutions (Left) Cell cytoplasm and solution have the same concentration of water molecules (Centre) A plasmolysed cell cytoplasm has a higher concentration of water molecules than the solution (Right) A turgid cell – the solution has a higher concentration of water molecules than the cell cytoplasm

Experiments to show osmosis

Non-living systems

In the apparatus in Fig. 2d.07, visking tubing is partially permeable and is used as a model for a cell membrane. The visking tubing bag is attached to a glass tube.

The solution inside the tubing contains sucrose (not glucose).

After a while the level of liquid in the tube rises, showing that water is entering the visking tubing due to osmosis.

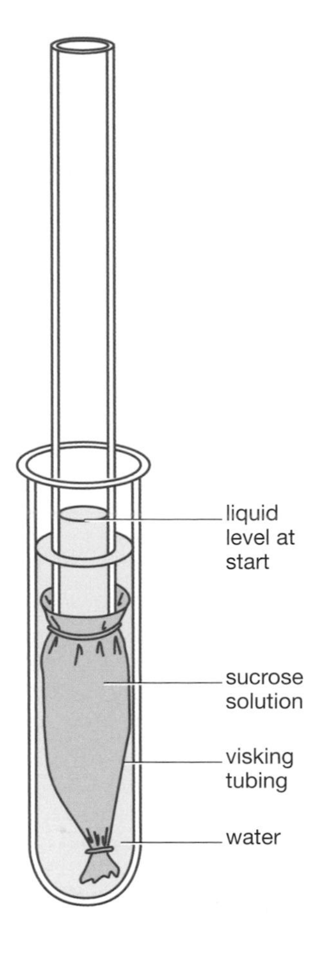

Fig. 2d.07: Apparatus for osmosis experiment

Living systems

- Measuring the length and/or mass of potato chips before and after being placed in solutions of different concentration will show a gain in length/mass when in solutions of higher concentration of water molecules, and loss of length/mass when in solutions of lower concentration of water molecules.
- Placing solutions of different concentration in carved-out potatoes, will show water lost from potato cells when the solution has a lower concentration of water molecules and water gained by potato cells when the solution has a higher concentration of water molecules.
- Comparing plant cells, e.g. beetroot or red onion cells, under a microscope after being placed in solutions of different concentration.
- If the plant cell is surrounded by a solution more dilute than its cytoplasm, the cell will absorb water by osmosis and become turgid (high internal pressure due to increased volume).
- If the cell is surrounded by a solution that is more concentrated than its cytoplasm, it will lose water by osmosis and become flaccid (low internal pressure due to decreased volume – if all the cells in a plant are flaccid, the plant wilts).
- If the volume of the plant cell shrinks too much, its cell membrane will shrink from the cell wall, and the cell may become a round circle. A cell like this is plasmolysed.

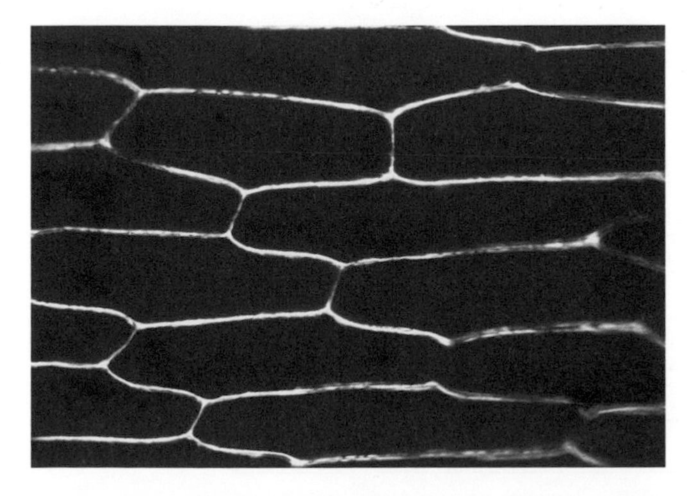

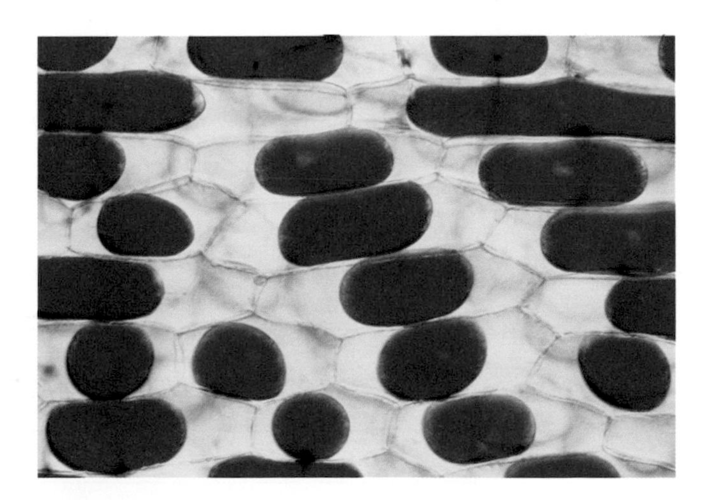

Fig. 2d.08: (Left) The cells have been in a solution of high concentration of water molecules and are turgid (Right) The cells have been in a solution of low concentration of water molecules and have lost water from their cytoplasm (they are plasmolysed)

Active transport

Active transport occurs when energy from respiration is used to take a molecule through a partially permeable membrane *against the molecule's concentration gradient*.

- It is an *active* process, not passive.
- The energy for active transport comes from respiration.
- Proteins in the cell membrane control active transport.

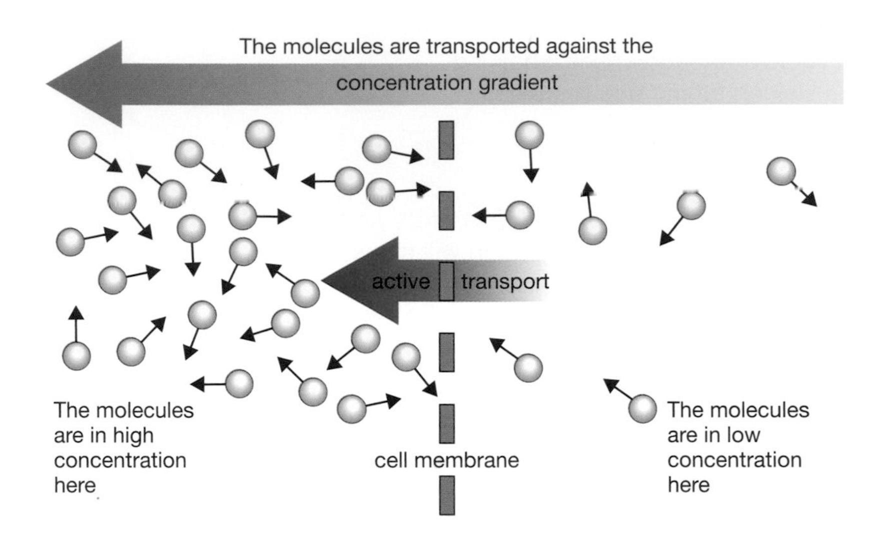

Fig. 2d.09: Active transport

CAM

Examples of active transport

- The concentration of ions in the soil water surrounding roots is at a lower concentration than inside the root cells. Root hair cells use active transport to take in essential ions such as magnesium and nitrate.
- In the small intestine, glucose crosses the cell membranes into villi cells by active transport.

Factors that affect the rate of movement into and out of cells

Surface area to volume ratio

This ratio (SA:V) is the surface area of the organism divided by its volume, e.g. x cm^2 divided by y cm^3.

Length of side (cm)	Surface area (cm^2)	Volume (cm^3)	Surface area to volume ratio
1	6	1	6 : 1 = 6
2	24	8	24 : 8 = 3
3	54	27	54 : 27 = 2

This is important for organisms. Larger organisms have a smaller SA:V ratio than smaller organisms.

- The volume of the organism is the number of cells that need to exchange materials with the environment.
- The surface area is the area over which the materials are exchanged.

So, a smaller SA:V ratio makes it more difficult for the cells to exchange as much of the materials as they need as quickly as they need them.

The rate of exchange across a surface can be increased by increasing the surface area, e.g. the alveoli in lungs, mesophyll cells in a leaf, root hair cells in plant roots, villi in the small intestine.

Temperature

An increase in temperature increases the movement (kinetic energy) of molecules. This increases the rate of diffusion or osmosis across a cell membrane.

As active transport involves proteins, a high temperature will denature the protein and active transport will stop.

Concentration gradient

The steeper the concentration gradient, the faster diffusion and osmosis occur.

The less steep the concentration gradient, the slower the rate of diffusion and osmosis.

You should now be able to:

★ define *diffusion*, *osmosis* and *active transport* (see pages 33, 35 and 38)
★ give examples where diffusion in gases and solutes is important for living organisms (see page 34)
★ describe how glucose enters and leaves a cell by diffusion (see page 34)
★ describe how water enters and leaves a cell by osmosis (see page 35)
★ describe the role of turgid cells in the support of plants (see page 36)
★ describe the effect of osmosis on an animal cell, such as a red blood cell (see page 36)
★ describe how cells take in substances by active transport (see page 38)
CAM ★ give one example of active transport (a) in plants, (b) in animals (see page 38)
★ describe how the rate of movement of substances into and out of cells is affected by (a) surface area to volume ratio, (b) temperature and (c) concentration gradient (see pages 38–39)
★ describe simple experiments to show diffusion (a) in living tissue, (b) in a non-living system (see page 35)
★ describe simple experiments to show osmosis (a) in living tissue, (b) in a non-living system (see page 37).

Practice questions

1. (a) Is a cell membrane permeable, impermeable or partially permeable? **(1)**

   (b) Is a cell wall permeable, impermeable or partially permeable? **(1)**

   (c) What part of a cell is required for active transport to occur? **(1)**

   (d) What kind of membrane is required for osmosis to occur? **(1)**

   (e) Haemoglobin is in red blood cells. Pure haemoglobin is soluble and forms a clear red solution. Red blood cells make a salt solution look cloudy red when the salt solution is the same concentration as the red blood cells' cytoplasm. But when red blood cells are placed in very dilute solutions, the solution becomes clear red. Explain what has happened. **(4)**

   (f) What would happen to a plant cell placed in a very dilute solution? **(1)**

   (g) A piece of fresh potato placed in salty water for one hour becomes floppy, shrinks, and tastes salty. Explain these observations. **(3)**

E Nutrition

You will be expected to:

CAM ★ define *nutrition*

Flowering plants

- ★ describe the process of photosynthesis and give the word and symbol equations for the reaction
- ★ describe the importance of photosynthesis in converting light energy to chemical energy
- ★ describe how carbon dioxide concentration, light intensity and temperature affect the rate of photosynthesis
- CAM ★ define *limiting factor* and explain it in terms of photosynthesis
- ★ describe simple experiments to investigate factors in photosynthesis
- ★ explain how a leaf is adapted for photosynthesis
- ★ identify some mineral ions that plants need for healthy growth
- CAM ★ explain the effects of nitrate ion and magnesium ion deficiency in plants

Humans

- ★ describe a balanced diet
- ★ identify sources, and describe functions, of a range of substances in the diet
- CAM ★ describe the uses, benefits and risks of food additives
- ★ describe how energy requirements vary in different groups of people
- CAM ★ describe the effects of malnutrition
- ★ identify structures of the human alimentary canal
- CAM ★ describe the structure and function of different types of human teeth, and explain why teeth need proper care
- ★ describe *ingestion, digestion, absorption, assimilation* and *egestion*
- ★ explain the role of peristalsis in the gut
- ★ describe how enzymes digest starch, proteins and lipids
- ★ describe the production of bile and its role in digestion
- ★ explain how the villus is adapted for absorption
- ★ describe an experiment to determine the energy content of a sample of food
- CAM ★ describe the role of the liver in metabolism, including deamination.

Nutrition and nutrients

Nutrition is the taking in, absorption and assimilation of nutrients by an organism. Nutrients include:

- organic substances from plant and animal tissue
- mineral ions.

Nutrients contain the raw materials or energy for growth and tissue repair.

Photosynthesis

Plants contain **chlorophyll**, a green pigment that absorbs light energy. Cells convert this energy to chemical energy in the form of chemical bonds in a simple sugar, usually glucose, built from carbon dioxide and water. The glucose is then used directly by the cells in respiration or converted to other biological molecules. Examples are lipids, amino acids, DNA, chlorophyll, sucrose, starch and cellulose. Once the energy is in the form of chemical energy, it can be taken in by animals for their nutrition.

Word equation for photosynthesis:

$$\text{carbon dioxide + water} \xrightarrow[\text{chlorophyll}]{\text{light energy}} \text{sugar (glucose) + oxygen}$$

Balanced equation for photosynthesis:

$$6CO_2 + 6H_2O \xrightarrow[\text{chlorophyll}]{\text{light energy}} C_6H_{12}O_6 + 6O_2$$

Factors that affect photosynthesis

The rate of photosynthesis can be changed if the following factors change.

- Carbon dioxide: The rate of photosynthesis increases with increasing carbon dioxide up to a point beyond which the rate levels out. (See Fig. 2e.01a.)

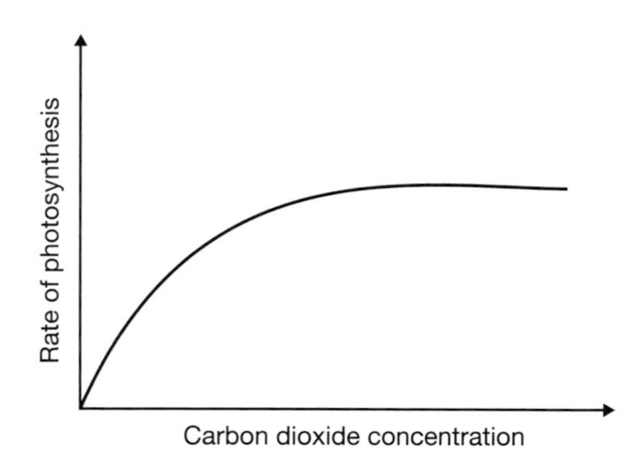

Fig. 2e.01a: Rate of photosynthesis with carbon dioxide

- Light intensity: As the light gets brighter, the rate of photosynthesis increases up to a point beyond which the rate levels out. (See Fig. 2e.01b.)

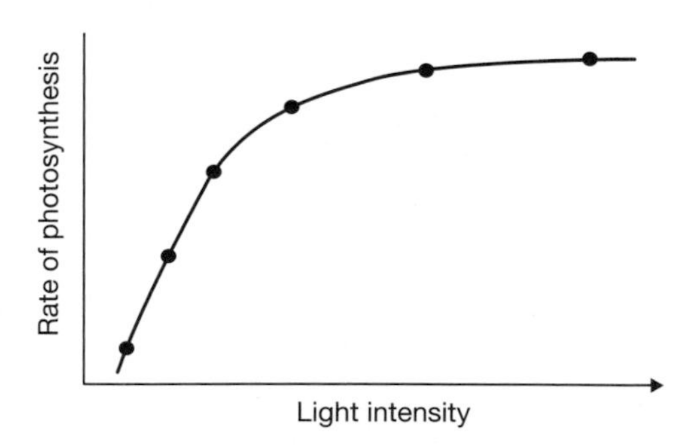

Fig. 2e.01b: Rate of photosynthesis with light

- Temperature: The rate of photosynthesis increases with increasing temperature up to a maximum of between 25–30 °C (for most plants). It then falls due to denaturation of the enzymes that control photosynthesis. (See Fig. 2e.01c.)

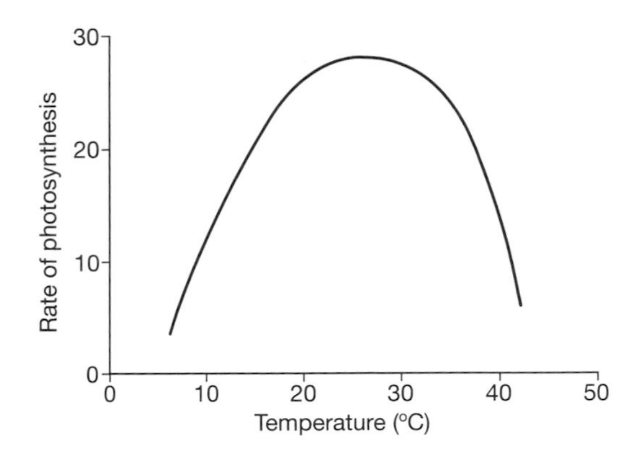

Fig. 2e.01c: Rate of photosynthesis with temperature

CAM

Limiting factors

In any particular set of conditions, the factor that is in shortest supply controls the rate of photosynthesis. We call this the **limiting factor**.

If that factor increases, the rate of photosynthesis will increase until another factor limits the rate and becomes the limiting factor, as shown in the graph.

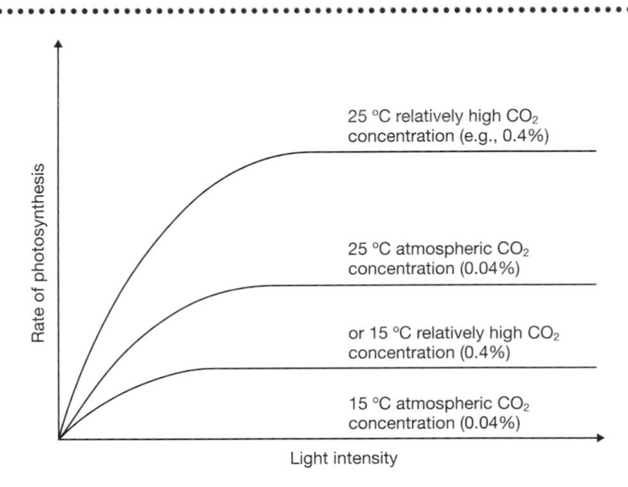

Fig. 2e.02: Limiting factors

Leaf structure and photosynthesis

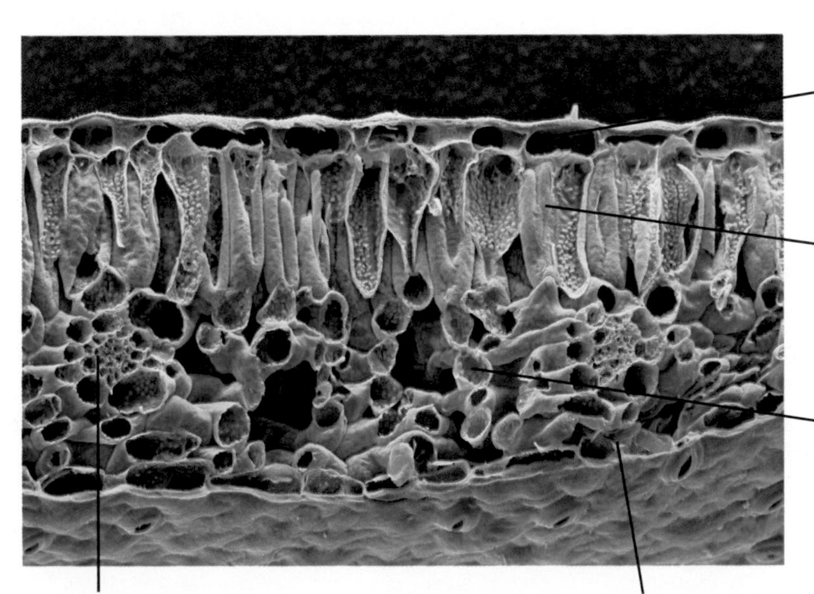

Fig. 2e.03: Leaf section

Experiments to investigate photosynthesis

Evolution of oxygen in light conditions and carbon dioxide in dark conditions

- 3 boiling tubes labelled 1, 2 and 3.
- Put a piece of pondweed into 1 and 2.
- Fill 1–3 with bicarbonate indicator.
- Tube 3 is the control.
- Put tube 1 in a dark cupboard.
- Put tube 2 near a cool fluorescent light.
- Leave for a minimum of 45 minutes or overnight.
- Tube 1 should go orange or yellow due to CO_2 from respiration lowering the pH (more acidic).
- Tube 2 should turn purple due to O_2 from photosynthesis raising the pH (more alkaline).

Starch production

Sugars produced by photosynthesis are converted to starch for storage. So the presence of starch in a leaf indicates that photosynthesis has taken place.

- Remove the leaf from a plant and put it in hot ethanol to remove the chlorophyll and soften the tissue.
- Rinse the leaf in cold water.
- Spread the leaf on a white tile and test with iodine solution.
- A blue-black colour shows the presence of starch.

Light requirement

The plant for testing must first be **de-starched**, by placing it in the dark for 1–2 days.

- Test a leaf for presence of starch to show that all starch has gone.
- Cover part of a leaf with black paper or metal foil.
- Leave the plant in the light for at least an hour or overnight.
- Prepare the leaf by putting it in hot ethanol, rinsing it in cold water, and testing it for starch (see Starch production above) – starch should have been formed only on the uncovered parts of the leaf, showing light is needed for photosynthesis.

Carbon dioxide requirement

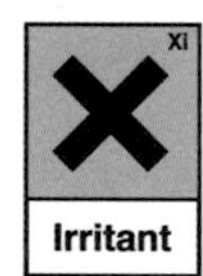

De-starch two plants by placing them in the dark for 1–2 days.

- Place each plant in a clear plastic bag (or under a glass bell jar).
- In one bag place a small dish of marble chips and dilute hydrochloric acid (to generate carbon dioxide).
- In the other bag add a beaker of potassium hydroxide solution (to remove carbon dioxide from air).
- Leave the plants for 24 hours. Prepare one leaf from each plant by putting it in hot ethanol, rinsing it in cold water, and testing it for starch (see Starch production above).
- The leaf with added carbon dioxide should have formed starch, the leaf without should not have formed starch, showing that carbon dioxide is needed for photosynthesis.

Chlorophyll requirement

Take a leaf from a plant that has variegated leaves (has green and white markings) and that has been left in the light overnight. The white areas have no chlorophyll.

- Prepare the leaf in hot ethanol and cold water, and test it for starch.
- Only the parts that were green should test positive for starch, showing that chlorophyll is needed for photosynthesis.

Mineral nutrition of plants

Photosynthesis only produces carbohydrates. Other nutrients are needed to make other biological molecules:

- magnesium is needed to make chlorophyll
- nitrogen is needed to make amino acids (for proteins).

Plants absorb these ions through their roots from the soil water.

CAM

Ion deficiency in plants

Since plants need nitrogen to make amino acids, and so make proteins, a deficiency of nitrogen means new cells cannot be formed, and other essential proteins cannot be made. Nitrate deficiency results in stunted growth.

Fig. 2e.04: Leaves from a nitrogen-deficient plant

Fig. 2e.05: Leaves from a magnesium-deficient plant

If a plant is magnesium-deficient, it cannot make chlorophyll. The leaves turn yellow and growth will be limited because the plant cannot make food.

Human diet

Our diet must supply all the materials we need to carry out life functions, including molecules used as the fuel for respiration and the raw materials to synthesise all the biological molecules in our bodies. A diet with the right amounts of each material is called a **balanced diet**.

The main food groups that we need are shown in the table.

Food group	Sources	Main uses	Notes
carbohydrates	bread, pasta, rice, potatoes, pulses (e.g. beans)	fuel for respiration	some stored as glycogen in liver and muscle cells
protein	meat, eggs, milk products, pulses (e.g. beans)	protein synthesis for growth and repair of cells	excess is converted into urea by liver and urea is excreted by kidneys
lipids (fats and oils)	vegetable oil, animal fat, red meat, full-fat milk products, nuts	fuel for respiration, some hormones and cell structures	excess stored as fat under skin and around some organs

We also need small amounts of **vitamins** and **minerals**, which play essential roles in some processes in the body.

Vitamins	Sources	Main uses	Deficiency symptoms CAM
vitamin A	deep green and orange vegetables and fruits, egg yolks, milk products	• helps prevent damage to cell membranes • used to make the light-sensitive pigment in retinal cells in the eye	poor night vision
vitamin C	citrus fruits, leafy green vegetables	• keeps tissues, including hair and skin, healthy as it is an antioxidant • helps in the absorption of iron	scurvy, where gums bleed and teeth fall out
vitamin D	milk products, oily fish and egg yolk made by the skin in sunshine	• helps in the absorption of calcium, so helps to keep bones and teeth strong	rickets, where bones are too soft and bend as they grow
Minerals			
calcium	milk products, dark green vegetables, pulses (beans)	• needed for keeping bones strong, in nerve impulse transmission, muscle contraction and blood clotting	muscle weakness, stunted growth
iron	leafy green vegetables, meat, egg yolks, whole grains, pulses	• required for the synthesis of haemoglobin in red blood cells	anaemia (blood lacking in haemoglobin), fatigue

In addition to the nutrients on the previous page, our diet needs to include plenty of:

- water because many of our metabolic reactions in cells occur in solution. It is also needed in the blood for transporting cells and solutes around the body, and for making certain secretions (sweat, digestive juices, mucus, tears and milk).
- **fibre** (roughage) from the cell walls of fruits, vegetables and whole grains. We do not absorb it, but it is important because it is attracted to water. This keeps the contents of the alimentary canal bulky and helps move food along. A lack of fibre in the diet is associated with constipation and diseases such as bowel cancer.

Food additives

We add many substances to food when we prepare it. These **food additives** include:

- preservatives that reduce the rate of growth of microorganisms, such as high concentrations of salt (e.g. salted meats), sugar (e.g. jams), vinegar (e.g. pickles)
- emulsifiers, to keep fats and oils mixed with the water-based part of a food (e.g. in ice cream)
- sweeteners, such as sugar or artificial sweeteners
- colours, to enhance colour especially after cooking
- thickening agents, such as starch (in sauces) or pectin (in jams).

The benefits of using additives are that they:

- prevent food decay, so food lasts longer (preservatives) and the risk of food poisoning is reduced
- make food more palatable (emulsifier, thickener, salt, sugar) or attractive (colour).

Some additives are associated with health hazards:

- Too much salt in the diet may cause high blood pressure in some people.
- Too much sugar in the diet can lead to obesity.
- Some artificial colours (e.g. tartrazine and sunset yellow) are linked to hyperactivity in children.

Energy requirements

The diet also supplies energy for respiration.

The energy required varies for different people:

- Active people (higher rate of metabolism) require more energy from food each day to maintain the same weight than those who are less active (lower rate of metabolism). This relationship is shown in the bar chart below.

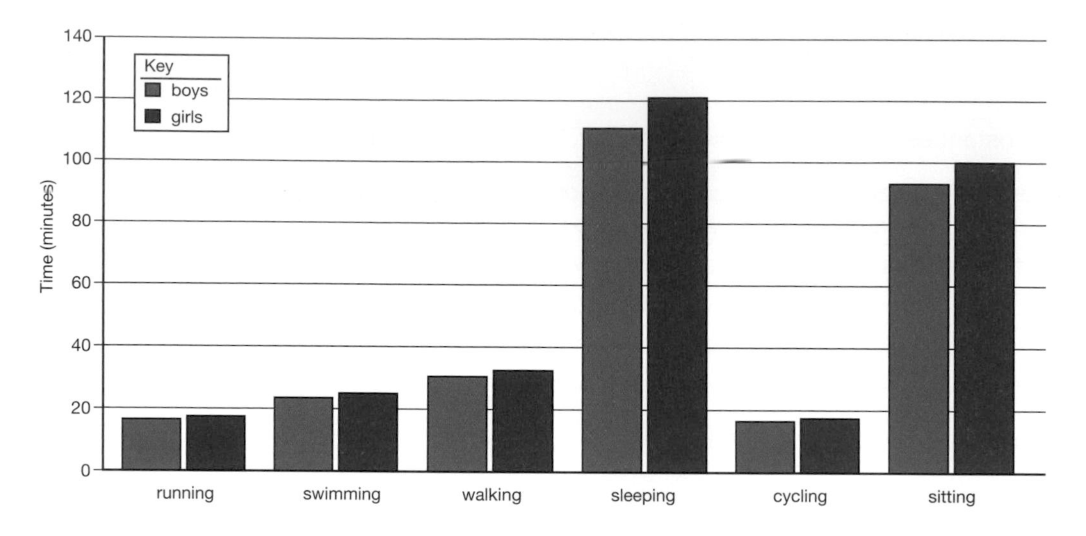

Fig. 2e.06: The number of minutes taken to use 400 kJ of energy

- The energy requirement for growth increases from birth up to the age of 15–18 years, when the rate of growth slows down a great deal; from about the age of 50, the metabolic rate starts to fall.
- The energy requirement during the last three months of pregnancy is higher (+800 kJ per day) due to the increased growth rate of the fetus and of some of the mother's tissues.

Average kJ required per day

Age	Male	Female
0–3 months	2280	2160
4–6 months	2890	2690
7–9 months	3440	3200
10–12 months	3850	3610
1–3 years	5150	4860
4–6 years	7160	6460
7–10 years	8240	7280
11–14 years	9270	7720
15–18 years	11 510	8110
19–50 years	10 600	8110
51–59 years	10 600	8000
60–64	9930	7990
65–74 years	9710	7960
75+ years	8770	7610

Malnutrition

Malnutrition is the condition that results from not getting the right kind of diet. It results in health problems, and can be caused by too much or too little of some factors in the diet.

- **Starvation** is not getting enough food. This means that many nutrients are in limited supply to the body. This can cause deficiency symptoms for vitamins and minerals (see table above). Too little energy in the food causes the body to break down food reserves – initially fat, but then muscle (protein) tissue – which can damage the heart and other tissues.
- Too much energy in the food (beyond what is needed by the body for processes such as growth and repair, and for exercise) causes fat to be laid down under the skin and around organs. A lot of fat results in **obesity** (being very overweight). Obesity is linked to heart disease, diabetes and other diseases.
- Too much saturated fat (from meat and dairy products) in food is linked to increasing deposits of cholesterol on the inside walls of arteries. This narrows the blood vessels, increasing blood pressure and reducing blood flow through the vessels. If the blocked artery supplies blood to the heart muscle, this can lead to a heart attack.
- Too little fibre in the diet can make it more difficult for peristalsis to move food along the alimentary canal. This can make faeces hard and dry and difficult to egest, which is **constipation**. It may also increase the risk of bowel cancer.

Structures of the alimentary canal

The main structures of the human alimentary canal are the mouth, oesophagus, stomach, small intestine, large intestine and pancreas.

CAM

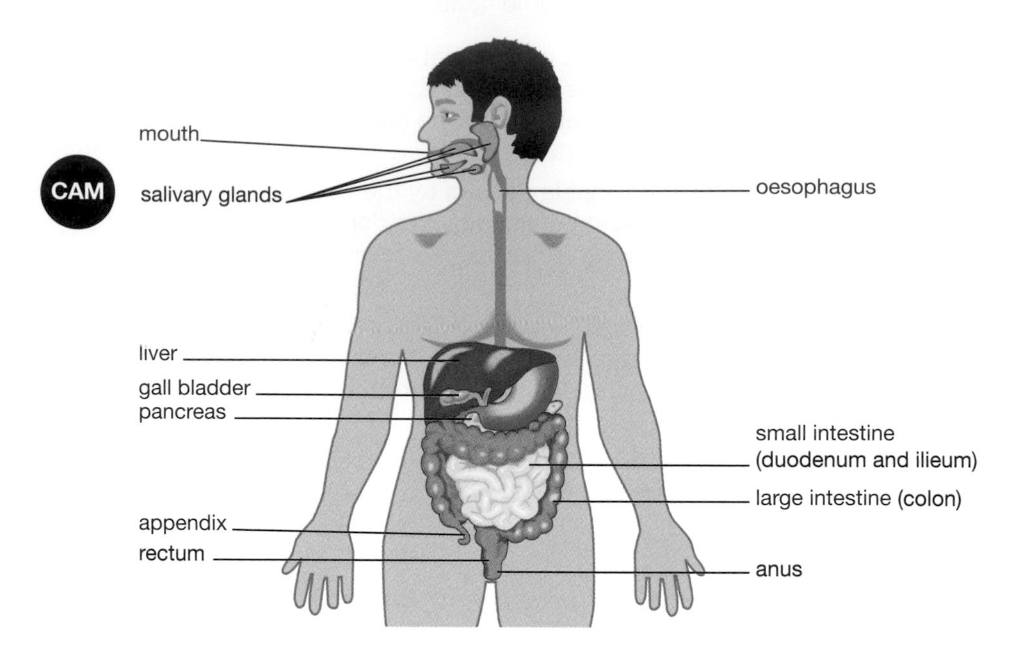

Fig. 2e.07: Structures in the human alimentary canal

The main functions of these structures are shown in the table.

Structure	Function	Notes
mouth	• where food is **ingested** • teeth break up food by chewing • saliva contains amylase which starts the chemical digestion of starch	chewing is mechanical digestion
oesophagus	• tube along which food passes from mouth to stomach, by **peristalsis**	vomiting is when peristalsis occurs in the opposite direction
stomach	• acid and protease enzymes secreted • food mixed with secretions by muscular wall of stomach	stomach enzymes work best at low pH (c. pH 2)
small intestine (duodenum and ileum)	• bile added to help neutralise acid from stomach • pancreatic secretions enter first part of small intestine (duodenum), include enzymes to complete chemical digestion and alkaline secretion to make food mix • slightly alkaline • specialised cells lining the small intestine (duodenum) secrete digestive enzymes • digested food molecules absorbed in rest of small intestine (ileum) • intestine wall has large surface area due to **villi** • a lot of water is absorbed into the body in the ileum also	enzymes here work best at a higher pH (c. pH 8) enzymes are amylase and other carbohydrases (enzymes that break down carbohydrates), proteases and lipases

CAM

CAM

Structure	Function	Notes
large intestine (colon)	• some water absorbed here, leaving **faeces**	friendly bacteria such as *E. coli* live in the large intestine and produce vitamin K which is important in blood clotting
pancreas	• gland that secretes many digestive enzymes, and alkaline secretions	enzymes include proteases, lipases and amylases
liver	• produces bile, which is alkaline, and it emulsifies lipids • assimilates many food molecules	
gall bladder	• stores bile from the liver until needed for digestion	
salivary glands	• amylase is secreted in the slightly alkaline saliva	saliva is very watery and contains mucus so it lubricates our food in the mouth before swallowing
rectum	• storage of faeces until ready for egestion	
anus	• opening of alimentary canal through which faeces are egested	

Processes in the alimentary canal

- **Ingestion:** eating, the taking in of food and drink into the alimentary canal through the mouth.
- **Digestion:** the breakdown of the large, insoluble biological molecules of carbohydrates, proteins and lipids into small soluble molecules that can pass through the alimentary canal wall.
- **Absorption:** the diffusion or active transport of small soluble molecules through the cells of the alimentary canal wall into the blood.
- **Assimilation:** the uptake of digested food molecules by cells from the blood so that the cells can synthesise the carbohydrates, lipids and proteins they need.
- **Egestion:** the elimination of undigested material from the alimentary canal.

Make sure you know the difference between **egestion** of undigested food from the alimentary canal and **excretion** of substances made in the body, such as urea from the kidneys and carbon dioxide from the lungs. Egestion and excretion are commonly confused.

Peristalsis

The wall of the different sections of the alimentary canal contains longitudinal and circular muscles.

Peristalsis is the succession of waves of contraction and relaxation of these muscles that pushes the food so that it can only go in one direction.

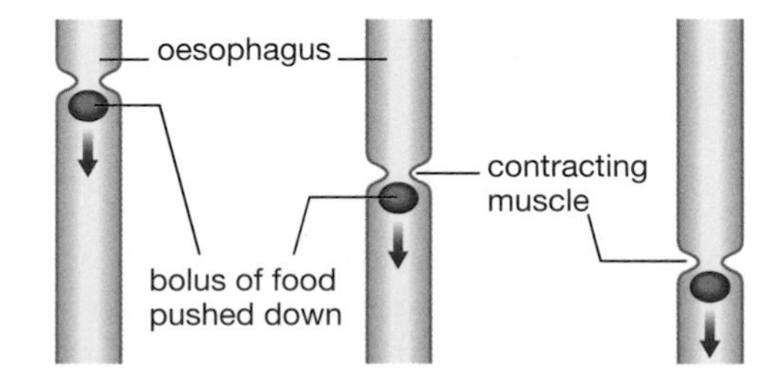

Fig. 2e.08: Peristalsis moves food through the alimentary canal

The role of digestive enzymes

Digestive enzymes result in chemical digestion in the alimentary canal of large insoluble food molecules into smaller soluble molecules that can be absorbed into the body.

Nutrient	Enzymes	Source	Site of action	Digests ...
carbohydrate	amylase	• salivary glands • pancreas	mouth small intestine	starch to maltose (not absorbed)
	maltase	• cells of small intestine wall	small intestine	maltose to glucose
protein	proteases	• cells of stomach wall • cells of small intestine wall • pancreas	stomach small intestine	proteins digested bit by bit to amino acids
lipid	lipases	• cells of small intestine wall • pancreas	small intestine	lipids to fatty acids and glycerol

CAM

Teeth

Food is broken into smaller pieces by **mechanical digestion** in the mouth, caused by biting and chewing with the teeth and the movement of the tongue.

Different types of teeth in the mouth have different purposes:

- **incisors** have a blade-like edge for biting off pieces of food
- **canines** are pointed and hold food while it is being bitten off
- **premolars** and **molars** have grinding surfaces for chewing.

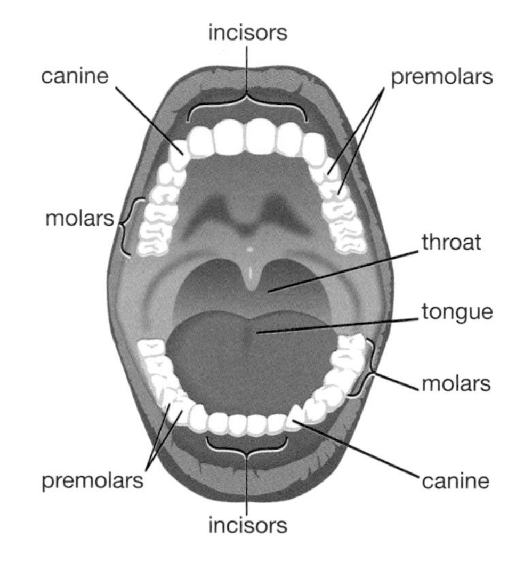

Fig. 2e.09: Teeth in the human mouth

Tooth decay

Bacteria in the mouth produce acids when they digest food. Food particles caught between the teeth encourage bacterial growth, and so acid is produced next to the teeth. The acid can dissolve the enamel that surrounds and protects the tooth, leading to **tooth decay.**

Tooth decay can be reduced by:

- brushing frequently to remove food particles left in the mouth after eating
- using a fluoride toothpaste, because fluoride strengthens enamel.

Fluoride is sometimes added to public water supplies to help protect everyone's teeth from tooth decay. However, in large amounts, fluoride can cause brown mottling of the teeth. Some people think it also causes health problems, but the evidence for this is variable. Some people think we should have the freedom to choose whether or not to take extra fluoride, rather than being given it in public water supplies.

Bile

Liver cells secrete **bile** continuously. Bile passes to the gall bladder where it is stored and concentrated.

- Bile is alkaline, so when it reaches the small intestine and mixes with the acidic contents from the stomach, a neutralisation reaction occurs. After neutralisation, any more bile makes the mixture slightly alkaline.
- Bile **emulsifies** lipids (fats and oils). This means it breaks the lipids up into much smaller globules. This increases the surface area of the lipids and so increases the rate of digestion by lipases.

Remember that lipids are insoluble in water and so are not easily broken down by water-soluble enzymes. Emulsification makes the globules of lipid much smaller, so that they mix more easily with the water-based contents of the small intestine.

Structure of a villus

The inner surface of the small intestine (ileum) is covered in millions of microscopic, finger-like structures called **villi** (singular, villus).

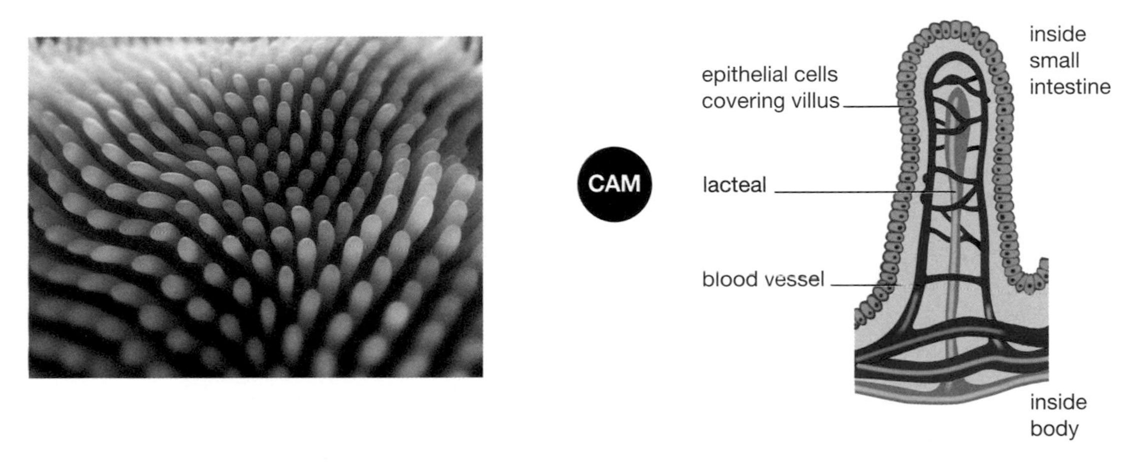

Fig. 2e.10: (Left) Many villi form the surface of the small intestine (Right) Structure of one villus

- Villi greatly increase the surface area of the small intestine, which increases the rate of diffusion, active transport and osmosis across cells of the small intestine wall.
- The epithelial cells on the surface of each villus have tiny projections called **microvilli** that increase the surface area even further.
- The surface of the villus is covered with a single layer of epithelial cells, so the distance that digested food molecules have to cross to the blood is as small as possible.
- Villi have a very good blood supply to carry away the products of digestion to the rest of the body.
- Each villus has a lacteal in its core, which transports lipids away from the small intestine, rather than in the blood. Lacteals are part of the lymphatic system. Lymphatic vessels carry lymph back to the heart, where the lymph re-enters the bloodstream. Lymph is clear and only has some white blood cells in it (lymphocytes) from the lymph nodes or glands that occur along the lymph vessels. After a very fatty meal, lymph turns milky white because of the fat droplets it has absorbed from the small intestine.

Assimilation in the liver

Digested food molecules absorbed into the blood in the small intestine are taken directly to the liver in the **hepatic portal vein.**

Many food molecules are assimilated by liver cells.

- Glucose is absorbed into liver cells and converted to glycogen for storage until more glucose is needed for respiration.
- Amino acids are converted to proteins.
- Excess amino acids (those that are more than the body needs) are broken down in liver cells. The nitrogen-containing part is combined with carbon dioxide to form **urea** in a process called **deamination**. The urea is excreted by the kidneys. The rest of the amino acid is used for respiration.

Liver cells also break down alcohol molecules and toxins, so that they cannot harm other cells and can be excreted from the body in urine.

Determining energy content in a food sample

Combustion (burning) releases the chemical energy in food as heat energy. Measuring the heat energy released can indicate the energy available from the respiration of the molecules in that food.

The apparatus for a simple experiment is shown in Fig. 2e.11.

- The food is set alight by placing it in a Bunsen burner flame.
- The burning food is then placed under the tube to heat the water, so that the heat energy is transferred to the water in the tube.
- The change in temperature of the water can be used to calculate the energy transferred to the water:

$$\text{energy transferred (J)} = \frac{(\text{temperature rise} \times 10)}{4.18}$$

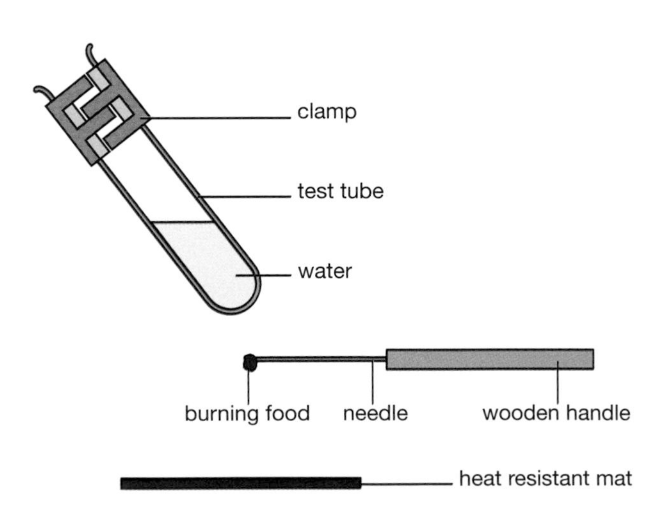

Fig. 2e.11: Combustion of food experiment

Note that this experimental set-up has many limitations, because only a proportion of the heat energy from the burning food is transferred to the water.

You should now be able to:

- CAM ★ explain what is meant by *nutrition* (see page 42)
- ★ describe the process of photosynthesis (see page 42)
- ★ state the word equation and the balanced chemical symbol equation for photosynthesis (see page 42)
- ★ describe how carbon dioxide concentration, light intensity and temperature affect the rate of photosynthesis (see pages 42–43)
- ★ explain how the structure of the leaf is adapted for photosynthesis (see page 43)
- ★ describe experiments to investigate photosynthesis, including:
  - the evolution of oxygen by a water plant
  - the production of starch
  - the need for light
  - the need for carbon dioxide
  - the need for chlorophyll (see pages 44–45)
- CAM ★ explain the term *limiting factor* in terms of photosynthesis (see page 43)
- ★ explain why plants need magnesium ions and nitrate ions (see page 45)
- ★ explain what is meant by a *balanced diet* (see page 46)
- ★ state the sources and describe the functions of the following in the diet: carbohydrates, proteins, lipids, vitamins, minerals, water, dietary fibre (see page 46)
- ★ describe how energy requirements vary with activity levels, age and pregnancy (see page 48)
- CAM ★ describe the uses, benefits and risks of food additives
- ★ describe the effects of malnutrition in relation to starvation, coronary heart disease, constipation and obesity (see page 49)
- ★ state the structures of the human alimentary canal and describe their main functions (see page 000)
- ★ explain what is meant by *ingestion*, *digestion*, *absorption*, *assimilation* and *egestion* (see page 000)
- ★ explain how and why food is moved through the gut by peristalsis (see page 51)

- ★ describe the structure and functions of the different types of human teeth (see page 52)
- ★ state the causes of dental decay and describe the proper care of teeth (see page 52)
- ★ explain arguments for and against the addition of fluoride to public water supplies (see page 52)
- ★ describe the role of amylase, maltase, proteases and lipases in the digestion of food, and state where these enzymes are produced in the alimentary canal (see page 52)
- ★ describe two functions of bile, and explain their importance (see page 53)
- ★ explain how the structure of a villus is adapted for absorption (see page 53)

- ★ describe the role of the liver in the metabolism of glucose, amino acids and toxins (see page 54).

Practice questions

1. Large insoluble food molecules are digested into smaller soluble food molecules that can pass into the gut.

   (a) Describe two ways that small soluble food particles are absorbed from the gut. **(2)**

   (b) Is water digested? Explain your answer. **(1)**

   (c) How is water absorbed from the gut? **(1)**

   (d) What do we call the large insoluble food molecules that we cannot digest? **(1)**

2. (a) Many foods contain fat. What are the two products of fat digestion? **(2)**

   (b) The liver produces a chemical substance called bile, which is added to food during digestion. Explain how bile helps in the digestion of fat. **(2)**

   (c) Describe one effect of eating more food than your level of activity requires and explain the effects this may have on the body. **(2)**

3.

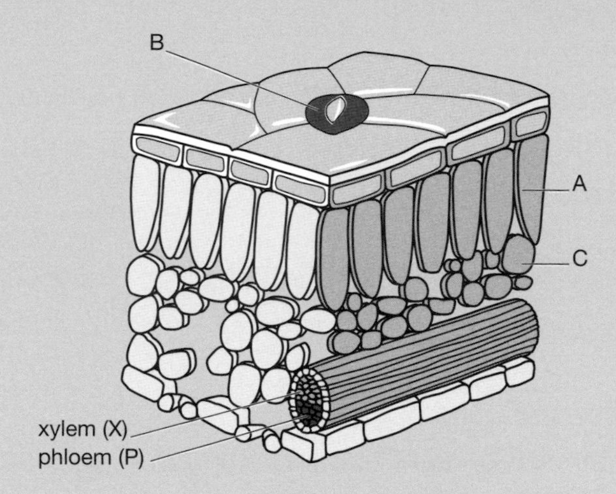

   (a) On the diagram of a leaf section above match each letter to the following functions:

   - Transports sugar
   - Allows carbon dioxide and oxygen in and out of the leaf
   - Carries water to the stem from the leaf
   - Absorbs the most light **(1)**

   (b) Within a UK rural area, the concentration of carbon dioxide is lowest around a cornfield at noon on a sunny day. What process is responsible for this? **(1)**

   (c) Within a UK rural area, the concentration of carbon dioxide is highest around a cornfield at midnight. What process is responsible for this? **(1)**

   (d) Name the green substance in the leaves of the corn. **(1)**

   (e) Where in the cells of the corn leaves is this substance found? **(1)**

   (f) What mineral from the soil is needed to make this green substance? **(1)**

F Respiration

You will be expected to:

- ★ state that respiration releases energy in living organisms
- CAM ★ state the uses of energy from respiration in the human body
- ★ describe the differences between aerobic and anaerobic respiration
- ★ state the word equation and balanced chemical symbol equation for aerobic respiration
- ★ state the word equation for anaerobic respiration in plants and animals
- CAM ★ state the balanced symbol equations for anaerobic respiration in animal cells, and in plant and fungal cells
- ★ describe a simple controlled experiment to demonstrate the evolution of carbon dioxide and heat during respiration by living organisms.

The process of respiration breaks down food molecules, such as glucose, into smaller molecules. The process releases energy that is used by the cell for all the life processes, such as growth and repair. Some of the energy is also transferred to the surroundings as heat energy.

CAM

In the human body, the energy from respiration is used for:

- the contraction of muscles
- the production (**synthesis**) of new molecules, such as proteins
- all the processes of cell division (see Section 3B)
- the active transport of molecules across membranes against their concentration gradient
- the passage of nerve impulses along nerves
- the maintenance of a constant core body temperature (see Section 2J).

Aerobic respiration

Aerobic respiration in living organisms uses oxygen to break down glucose into carbon dioxide and water. A lot of energy is released because many bonds are broken in each glucose molecule to form simpler carbon dioxide and water.

The word equation for aerobic respiration is:

glucose + oxygen → carbon dioxide + water (+ energy)

The balanced symbol equation for aerobic respiration is:

$$C_6H_{12}O_6 + 6O_2 \rightarrow 6CO_2 + 6H_2O \text{ (+ energy)}$$

Most of the chemical reactions for aerobic respiration occur in the mitochondria.

Anaerobic respiration

Anaerobic respiration is the breakdown of food molecules, such as glucose without the presence of oxygen. The glucose is only partly broken down so less energy is released from each glucose molecule than in aerobic respiration. Anaerobic respiration takes place in the cell cytoplasm.

Anaerobic respiration in animal cells

Anaerobic respiration takes place in muscle cells when not enough oxygen is available for aerobic respiration. The reaction produces lactic acid, and releases energy.

The word equation for this reaction is:

glucose → lactic acid (+energy)

The balanced symbol equation for anaerobic respiration in animal cells is:

$$C_6H_{12}O_6 \rightarrow 2C_3H_6O_3 \text{ (+ energy)}$$

The lactic acid produced accumulates in muscle cells and must be broken down after exercise. It is either converted back to glucose or used in respiration, which uses additional oxygen (sometimes called the 'oxygen debt').

The word equation for anaerobic respiration in plant and fungal cells is:

glucose → ethanol + carbon dioxide (+ energy)

Anaerobic respiration in fungi and plants

Anaerobic respiration takes place in plant cells when not enough oxygen is available, such as inside the food stores of seeds. The reaction produces ethanol and carbon dioxide and releases energy.

The balanced symbol equation for this reaction is:

$$C_6H_{12}O_6 \rightarrow 2C_2H_5OH + 2CO_2 \text{ (+ energy)}$$

Anaerobic respiration in fungi is more commonly called **fermentation**. This is made use of in:

- brewing beer, where the breakdown of sugars by yeast produces ethanol (alcohol)
- bread-making, where the carbon dioxide produced by yeast causes the dough to 'rise', making a softer, lighter bread than could be made without yeast.

Comparing aerobic and anaerobic respiration

	Aerobic respiration	Anaerobic respiration
substrate	glucose	glucose
oxygen needed	yes	no
products	carbon dioxide + water	(animals) lactic acid (plants/fungi) ethanol + carbon dioxide
amount of energy produced per molecule of substrate	much more than anaerobic	much less than aerobic
site	mitochondria	cytoplasm

Investigating respiration

We can use the apparatus shown in Fig. 2f.01 to investigate the evolution of carbon dioxide and heat from respiring seeds (peas).

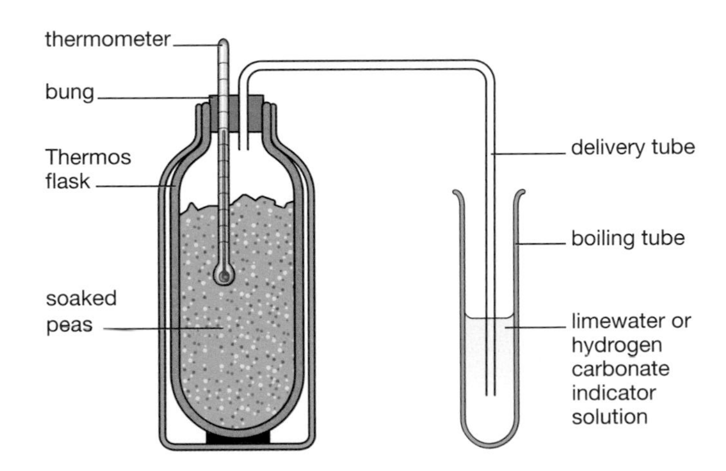

Fig. 2f.01: Experiment to investigate the evolution of carbon dioxide and heat from respiring seeds

- The peas are soaked in water to begin germination.
- Place the peas in a thermos flask, and fit a two-hole bung with a thermometer and delivery tube.
- The end of the delivery tube is placed into limewater or bicarbonate indicator, to react with any carbon dioxide.
- A control set-up uses the same apparatus but with boiled ('dead') peas.
- The temperature in the flask of 'live' peas should increase. Lime water turns cloudy and bicarbonate indicator turns yellow. This should not occur in the flask of 'dead' peas.

You should now be able to:

- ★ define the term *respiration* (see page 57)
- CAM ★ state the uses of energy in the body of humans (see page 57)
- ★ write the word equation and balanced symbol equation for aerobic respiration (see page 57)
- ★ write the word equations for anaerobic respiration (a) in animal cells, (b) in plant (and fungal) cells (see page 58)
- CAM ★ write the balanced symbol equation for anaerobic respiration (a) in animal cells, (b) in plant and fungal cells (see page 58)
- ★ draw up a table to compare aerobic and anaerobic respiration (see page 59)
- CAM ★ describe the role of anaerobic respiration in yeast during brewing and bread-making (see page 58)
- ★ describe simple controlled experiments to demonstrate the evolution of carbon dioxide and heat from respiration in living organisms (see page 59).

Practice questions

1. Usain Bolt broke the world record for the 100 m dash in 2009.

   (a) He did not appear to breathe during the race. How could his muscles still contract? **(2)**

   (b) Why was Usain Bolt breathing so heavily at the end of the race? **(2)**

   (c) Usain says he drinks one ginger beer a day. Ginger beer is a sweet, fizzy drink made using yeast. Ginger beer contains 1% ethanol. Write the word equation for how yeast can make ethanol. **(2)**

   (d) Give one example of how we use yeast. **(2)**

   (e) What chemical reaction do both Usain Bolt and bacteria in yoghurt use to release energy? **(2)**

   (f) Write a balanced equation for the above reaction. **(2)**

   (g) Plants can respire using oxygen. Write the balanced equation for this kind of respiration. **(2)**

G Gas exchange

You will be expected to:

★ describe the importance of diffusion in gas exchange

Flowering plants

★ explain the importance of gas exchange in plants in relation to respiration and photosynthesis

★ explain why net gas exchange changes in plants during the day and night

★ explain how the leaf is adapted for gas exchange

★ describe the role of stomata in gas exchange

★ describe simple experiments on net gas exchange from a leaf

Humans

★ describe the structures in the thorax used in ventilation

★ explain how the intercostal muscles and diaphragm produce ventilation

★ explain how alveoli are adapted for gas exchange

CAM ★ describe the differences between inspired and expired air

★ explain the role of mucus and cilia in protecting the lungs

★ describe the effects of smoking on the lungs and circulatory system

★ describe a simple experiment to investigate the effect of exercise on breathing

CAM ★ explain the effect of physical activity on breathing

Animals

CAM ★ describe how gas exchange structures in animals are adapted for diffusion.

Diffusion and gas exchange

Gases involved in respiration and photosynthesis enter and leave cells by diffusion. Factors that affect the rate of diffusion will also affect gas exchange between cells and the environment.

TIP

Apply what you learned about the effect of factors on the rate of diffusion in Section 2D, such as surface area to volume ratio, temperature and concentration gradient, to gas exchange in plants and animals in this section.

Gas exchange in flowering plants

Gas exchange in plants occurs to support respiration and photosynthesis.

- Respiration happens all the time – all cells take in oxygen and produce carbon dioxide.
- Photosynthesis occurs only in the presence of light – cells with chloroplasts take in carbon dioxide and produce oxygen.

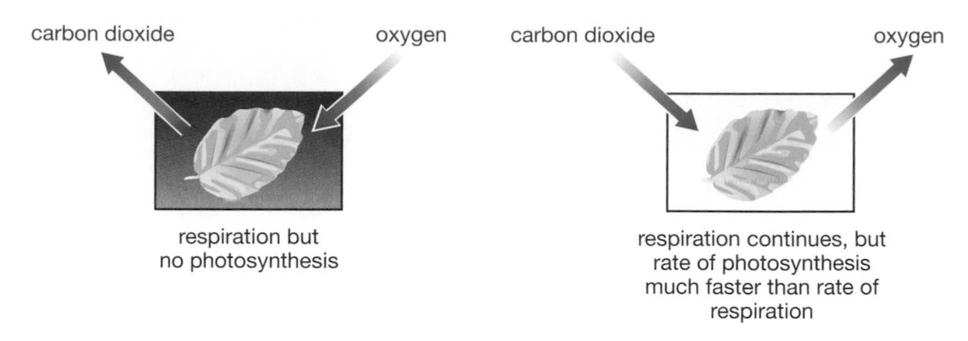

Fig. 2g.01: Net gas exchange

The rate of photosynthesis increases as light intensity increases. Respiration is not affected by light.

If we only measure the gases produced or taken in by a leaf, we measure the **net gas exchange** (see Fig. 2g.01).

- When it is dark, there is only respiration, so the net gas exchange is carbon dioxide out of the leaf and oxygen into the leaf.
- As light intensity increases, photosynthesis starts. The chloroplasts use carbon dioxide produced in respiration, and oxygen from photosynthesis is used by the plant in respiration. At a particular point the rate of carbon dioxide production from respiration balances the rate of use in photosynthesis, and the rate of oxygen production by photosynthesis balances the rate of use in respiration. The net gas exchange between the leaf and environment for both gases will be zero.
- At higher light intensities, the rate of photosynthesis exceeds the rate of respiration, so the net gas exchange is carbon dioxide into the leaf and oxygen out of the leaf.

Leaf adaptations for gas exchange

Stomata (see page 43) are pores in the leaf's epidermis through which carbon dioxide, oxygen and water vapour diffuse and are exchanged with the air outside the leaf.

Guard cells either side of a stoma cause the stoma to open when it is light, and close when the plant is wilting or it is dark.

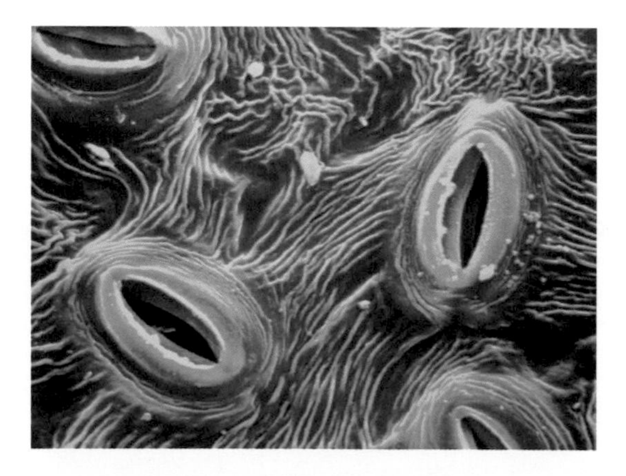

Fig. 2g.02: Stomata on the underside of a leaf

Investigating net gas exchange

- Place rinsed pondweed (e.g. *Elodea*) in two boiling tubes filled with water containing hydrogencarbonate indicator equilibrated with the air (red colour).
- Place one tube near a fluorescent (cool) lamp and the other in the dark.
- After 24 hours compare the colour of the indicator.

The plant in the dark will only respire and will produce carbon dioxide. The carbon dioxide will dissolve in the water and the indicator should turn yellow as the water becomes more acidic.

The water surrounding the plant in the light should turn purple, indicating more alkaline conditions as carbon dioxide is taken out of the water.

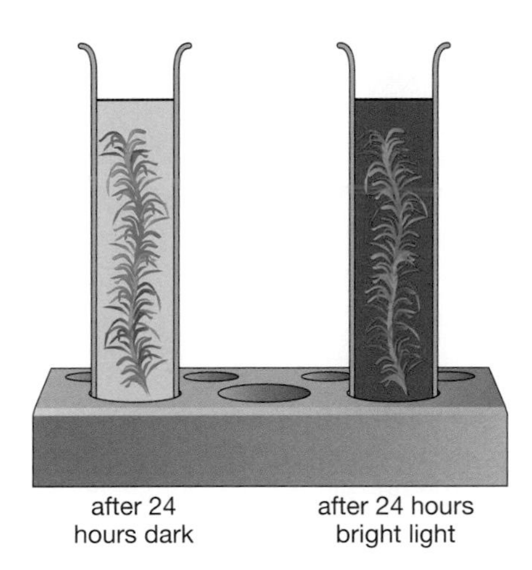

Fig. 2g.03: Pond weed in indicator solutions

CAM

Gas exchange in animals

Multicellular animals use a range of organs to increase the surface area for gas exchange.

- Mammals, reptiles, birds and adult amphibians have lungs.
- Fish and young amphibians have gills.
- Amphibians also use their soft wet skin for gas exchange.
- Many invertebrates use a form of gill, or breathing tubes that link the body to the environment.

All these organs have features that increase the rate of diffusion of gases which contribute to:

- maintaining a diffusion gradient for carbon dioxide and oxygen between the body and the environment, including ventilation movements and a good blood supply
- a large surface area over which diffusion can occur
- short diffusion pathways, often just two cells between the air or water and the blood.

Gas exchange in humans

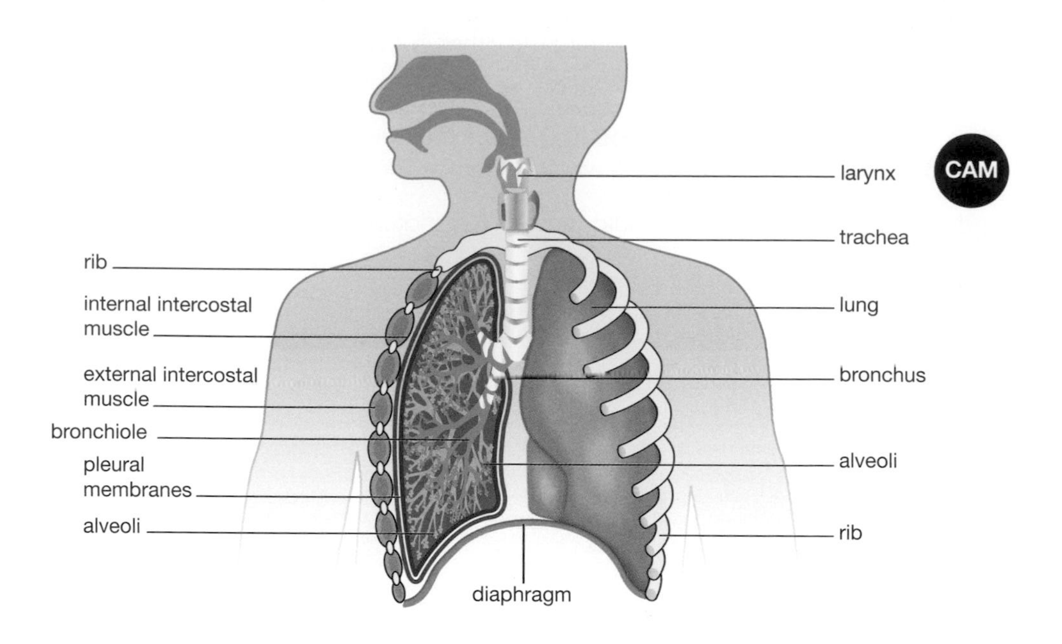

Fig. 2g.04: Structures in the thorax

The **thorax** contains the main structures in the human respiratory system:

- the **ribs** form a protective cage around the organs in the thorax – they can be moved by the intercostal muscles
- the **intercostal muscles** move the ribs during deep breathing (see below)
- the **diaphragm** is a sheet of tough tissue surrounded by muscle that separates the thorax from the abdomen, used in breathing

- the **larynx** at the top of the trachea contains vocal folds that produce sound as air passes over them which enable us to talk
- the **trachea** (windpipe) carries air from the mouth and nose to the lungs, supported by rings of cartilage to prevent collapse when breathing
- two **bronchi** (single, bronchus) join the trachea to the bronchioles – also supported by rings of cartilage
- the **bronchioles** are many-branched tubes leading to the alveoli, supported by rings of cartilage
- the **alveoli** are bubble-like sacs at the end of the bronchioles where gas exchange occurs – there are millions of these in each lung
- the **pleural membranes**, which surround and protect the lungs, secrete pleural fluid that makes it easier for the lungs to expand and contract without rubbing against the ribs.

Cells lining the bronchi and bronchioles secrete mucus to trap particles (e.g. dust and bacteria) that are in the air, so the particles do not reach the alveoli. The trachea also has a lining of ciliated cells which move the mucus up to the back of the throat, where it is swallowed and digested in the stomach.

Ventilation

Ventilation is the movement of air into and out of the lungs – commonly called breathing. Air enters the lungs during **inhalation**, and leaves during **exhalation**. Lungs contain elastic tissue. Lungs therefore expand during inhalation, and recoil during exhalation.

Inhalation

Inhalation is active and the intercostal muscles and the diaphragm muscle contract. Gentle breathing mainly involves the diaphragm pulling downwards on the thoracic cavity. When we take deeper breaths, the intercostal muscles also contract to increase the volume of the thorax even more.

When we inhale (breathe in):

- the muscle ring surrounding the diaphragm contracts and pulls it down
- the intercostal muscles (for a deep breath) contract and lift the rib cage up and out
- these movements increase the volume of the thorax which decreases the air pressure inside the lungs
- air pressure inside the lungs becomes lower than atmospheric air pressure outside the body
- air moves from the higher pressure to the lower pressure, so the outside air moves into the lungs.

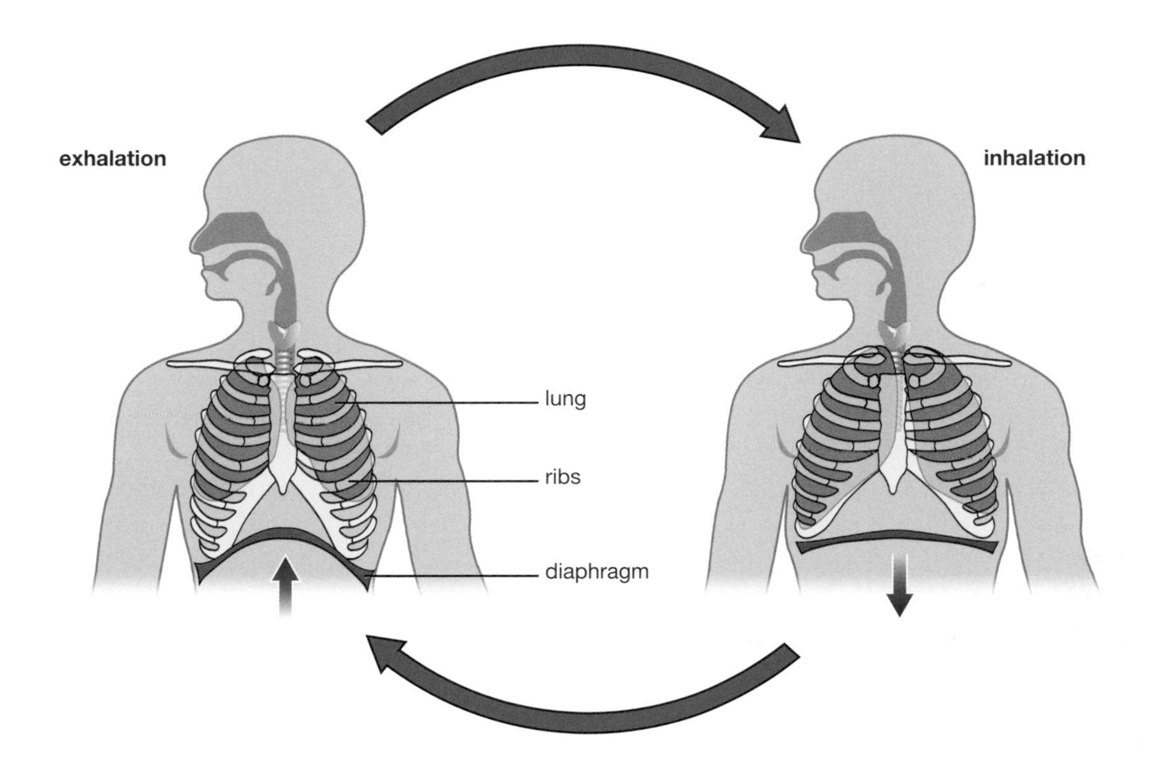

Fig. 2g.05: Exhalation and inhalation

Exhalation

Exhalation is passive.

When we exhale (breathe out):

- the diaphragm muscle relaxes and the diaphragm returns to its domed shape
- the intercostal muscles relax and the rib cage moves down and in
- both movements decrease the volume of the thorax which increases the air pressure inside the lungs
- air pressure inside the lungs becomes higher than atmospheric air pressure outside
- air moves from the higher pressure to the lower pressure, so air is pushed out of the lungs.

Inspired and expired air

As gases are exchanged in the lungs between the air and blood, the proportion of the gases are different in inspired air (from the environment) and expired air (from the body).

	Inspired air	Expired air
oxygen concentration	higher	lower
carbon dioxide concentration	lower	higher
water vapour concentration	usually lower	usually higher
temperature	lower	higher

We can show that carbon dioxide is in higher concentration in expired air by breathing through a straw (or tube) into a container containing limewater. The limewater turns cloudy, indicating carbon dioxide. Bubbling air through limewater should show little or no cloudiness, as the concentration of carbon dioxide in the air is relatively low.

Alveoli

Alveoli are adapted to maximise the rate of gas exchange by diffusion between the air in the lungs and the blood supply to the lungs in the following ways:

- they greatly increase the surface area of the lungs for diffusion
- each alveolus is in close contact with blood capillaries
- as the blood moves through the network of capillaries around each alveolus, it exchanges oxygen and carbon dioxide quickly – this maintains high concentration gradients for a rapid rate of diffusion
- the lining of an alveolus is one cell thick (0.2 μm wide), which provides a short diffusion distance.

The effects of smoking

Tobacco smoke contains tar, nicotine and carbon monoxide, which affect the respiratory system and the circulatory system.

- Chemicals in the smoke can paralyse the cilia lining the trachea, so mucus containing trapped bacteria stays in the lungs, and can collect in the bronchi, bronchioles and alveoli. The bacteria may multiply and infect these structures and can cause bronchitis and pneumonia.
- **Tar** in cigarette smoke is carcinogenic (can cause **cancer**). As the mucus in the lungs is not removed, the tar has more time to cause lung cancer.
- Coughing to remove sticky mucus damages the thin single-cell lining of the alveoli. The walls of the alveoli break down, reducing their surface area for gas exchange, and the remaining walls thicken, increasing the diffusion distance, slowing down gas exchange. This condition is called **emphysema**.
- Other chemicals in the smoke that diffuse into the blood are also carcinogenic. As they travel around the body they can cause cancers in many other tissues.

- **Nicotine** is highly addictive, making it very difficult to give up smoking. It increases alertness and suppresses the appetite. It raises the heart beat rate and the blood pressure. Withdrawal symptoms occur when a smoker tries to quit.
- **Carbon monoxide** forms in tobacco smoke. It is a poisonous gas because it prevents oxygen combining with haemoglobin. This reduces the amount of oxygen getting to tissues in the body, which can kill cells. In pregnant women, carbon monoxide from tobacco smoke means the developing fetus gets less oxygen. Its rate of growth slows down, causing low birth weight, which can be associated with many problems at birth and in later life.

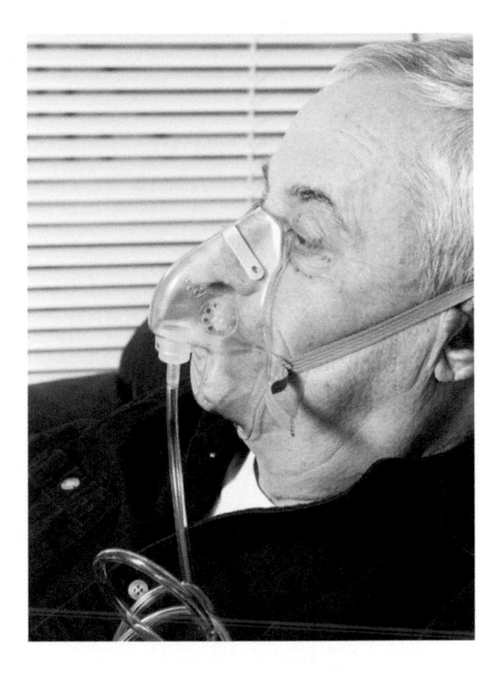

Fig. 2g.06: If the alveoli are very badly damaged, a patient may need to breathe gas that contains more oxygen than normal, so that enough oxygen can diffuse into the blood and be delivered to cells for aerobic respiration

Investigating the effect of exercise on breathing

The short-term effect of exercise on breathing can be investigated as follows.

- Take three readings of the breathing rate at rest from one individual and calculate the mean.
- Measure the breathing rate after 1 minute of gentle exercise, e.g. walking slowly. Allow the breathing rate to return to normal, and repeat twice more. Calculate the mean.
- Measure the breathing rate after 1 minute of vigorous exercise, e.g. step exercises. Allow the breathing rate to return to normal, and repeat twice more. Calculate the mean.

The results should show that increased levels of activity increase the breathing rate, because muscle cells are respiring more quickly to release energy for contraction. The rate of oxygen uptake by the muscle cells increases so the rate of delivery of oxygen to the muscles must increase. Carbon dioxide production in the muscle cells increases so an increased breathing rate helps to get rid of the carbon dioxide faster.

CAM

The effect of exercise on depth of breathing can be measured using a spirometer, or the displacement of water from a container by a normal breath measured after each level of exercise. This should show that depth of breathing increases with increased levels of exercise, to provide more oxygen more quickly to muscles and remove larger amounts of carbon dioxide more quickly. This prevents too much carbon dioxide dissolving in the blood and the cytoplasm of cells, reducing the pH and therefore affecting the rate of activity of enzymes.

The long-term effects of exercise on breathing rate can be investigated by comparing breathing rates at different levels of exercise for a group of trained athletes and a group of untrained healthy individuals.

You should now be able to:

- ★ describe the role of diffusion in gas exchange (see page 61)
- ★ describe and explain the net gas exchange of a plant during the day, and at night (see page 62)
- ★ explain how the structure of the leaf is adapted for gas exchange (see page 62).
- ★ describe a simple controlled experiment to investigate the effect of light on net gas exchange from a leaf (see page 63)
- CAM ★ describe how gas exchange structures in animals are adapted for diffusion (see page 63)
- ★ describe the structure and function of organs of the thorax in the respiratory system (see page 64)
- ★ explain how the intercostal muscles and the diaphragm work to produce ventilation of the lungs (see page 65)
- ★ explain the role of mucus and cilia in protecting the gas exchange system (see page 64)
- CAM ★ state the differences in composition between inspired and expired air (see page 66)
- ★ describe an investigation into the composition of inspired and expired air (see page 66)
- ★ explain how alveoli are adapted for gas exchange by diffusion (see page 66)
- ★ describe the effect of tobacco smoke on the lungs and the circulatory system (see pages 66–67)
- ★ describe a simple experiment to investigate the effect of exercise on the rate of breathing in humans (see page 67)
- CAM ★ describe a simple experiment to investigate the effect of exercise on the depth of breathing in humans (see page 67)
- ★ explain why rate and depth of breathing change in relation to level of activity (see page 67).

Practice questions

1. Air is full of particles that you do not want in your lungs.

(a) How is the trachea adapted to make sure the particles are removed? **(3)**

(b) Your lungs release a gas into the alveoli, which you breathe out. Explain how you breathe out. **(4)**

(c) Smokers deliberately inhale hot particles of tobacco smoke into their lungs. Describe two effects of smoking on the body. **(4)**

(d) Cartilage rings are found in which three structures in the lungs? **(3)**

(e) List the three adaptations of alveoli for efficient gas exchange. **(3)**

2. Leaves are adapted to carry out photosynthesis.

(a) Do the cells in leaves respire? **(1)**

(b) What gas is needed by the plant to photosynthesise? **(1)**

(c) By what passive process does this gas enter the leaf? **(1)**

(d) What are the structures in a leaf that control gas exchange with the environment? **(1)**

(e) Hydrogen carbonate indicator is red when normal air is dissolved in it, yellow when the air dissolved in it has more carbon dioxide, and purple when the air dissolved in it has no dissolved carbon dioxide. Describe a method, with controls, of how you could investigate gas exchange in the water plant *Elodea*. Labelled diagrams are acceptable. **(5)**

H Transport

You will be expected to:

- ★ explain why unicellular organisms can rely on diffusion but multicellular organisms need a transport system

Flowering plants

- ★ describe the role of phloem in transporting materials around a plant
- ★ describe the role of xylem in transporting water and mineral salts in a plant
- CAM ★ identify the positions of xylem and phloem cells in a dicotyledonous stem, root and leaf
- ★ explain how water is absorbed by root hair cells
- CAM ★ explain the pathway taken by water through root, stem and leaf cells
- ★ define the term *transpiration*
- CAM ★ explain the role of a water potential gradient and cohesion of water molecules in transpiration
- ★ explain how the rate of transpiration is affected by different factors
- ★ describe experiments on the effect of factors on the rate of transpiration
- CAM ★ define the terms *source* and *sink* in transpiration and translocation and how they change over the seasons
- ★ describe the translocation of systemic pesticides
- ★ compare the adaptations of garden plants, water plants and desert plants

Humans

- ★ state the composition of blood
- ★ describe the role of plasma in the transport of substances
- ★ describe how red blood cells are adapted to transport oxygen
- ★ describe how the immune system responds to infection
- ★ explain how vaccination protects the body from infection in the future
- ★ state the role of platelets in blood clotting
- CAM ★ describe the role of fibrinogen in blood clotting
- ★ describe the structure and function of the heart
- ★ describe how the heart rate changes during activity and in the presence of adrenaline
- ★ describe the main types of blood vessels and their roles
- ★ describe a general plan of the circulatory system
- ★ name the main blood vessels supplying the heart, lungs, liver and kidneys
- CAM ★ describe the effect of blockage of coronary arteries and suggest possible causes
- ★ describe the transfer of substances between capillaries and tissue fluid
- ★ describe the role of the lymphatic system.

The need for transport systems

Organisms need a system for transporting substances into their cells. Some organisms rely on diffusion for this but the organism's need for substances (dependent on its volume of tissue/cells) and its surface area may mean that diffusion alone is not enough, and that a bulk flow of materials around an organism to its cells is necessary. For example, it would take 70 years for an oxygen molecule to diffuse from your lungs to your big toe, but because you have a bulk flow of blood in your circulation system, the delivery of oxygen from your lungs to your big toe occurs in seconds.

As volume increases, the surface area to volume ratio decreases (see Section 2D), and it becomes more difficult for diffusion to supply all the needs of cells for efficient working.

- Small, unicellular organisms can rely on diffusion as they have a large surface area to volume ratio.
- Large organisms need a way of getting substances to and from cells faster than diffusion can manage. So multicellular organisms have **transport systems**, e.g. xylem and phloem in plants, circulatory systems in animals.

Transport in flowering plants

The veins (vascular bundles) in plants run throughout the stem, leaves and roots. They contain **xylem vessels** and **phloem tissue**.

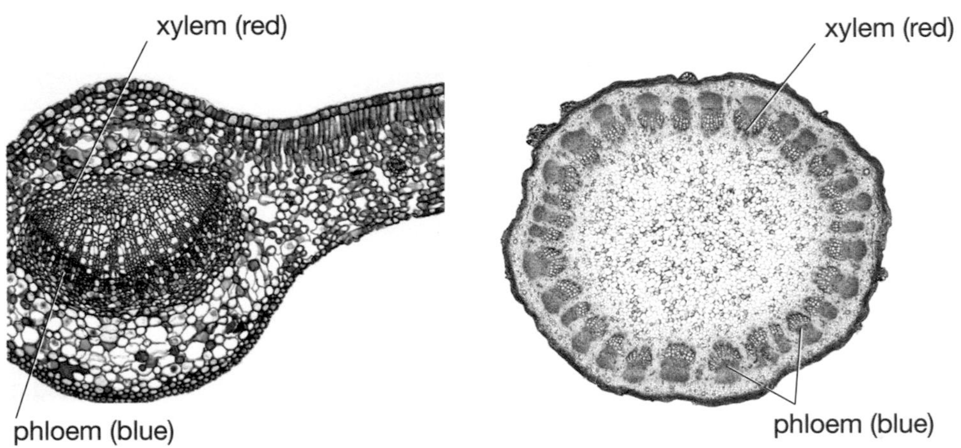

Fig. 2h.01: (Left) Cross section of a dicotyledonous leaf (Right) Cross section of a dicotyledonous stem

The structure of a dicotyledonous root, in cross section, shows an x-shaped xylem in the centre, with phloem tissue in the outer angles.

Phloem

Phloem comprises a bunch of separate phloem tubes. Each phloem tube is made up of living cells stacked one on top of the other with holes in their cross walls for continuity. These phloem tubes transport dissolved food molecules, e.g. sucrose and amino acids, up and down a plant.

- Sucrose is formed in leaves from glucose after photosynthesis. It is the carbohydrate transport molecule in plants, because sucrose is not as reactive as glucose. It is carried to other parts of the plant and converted back into glucose for respiration or conversion to starch for storage.
- Amino acids are made in leaves by combining carbohydrate products of photosynthesis with nitrogen supplied from the soil as nitrates. The amino acids are transported in the phloem to other parts of the plant, e.g. developing seeds.

Transport of dissolved food molecules may be up or down the separate phloem tubes, but in different cells. That is, one phloem tube may be transporting substances up and the phloem tube next to it may be transporting substances down.

The scientific term for the transport of substances such as sucrose and amino acids in the phloem is **translocation**.

- A **source** is the part of a plant that produces a substance or where a substance enters the plant.
- A **sink** is the part of the plant where a substance leaves the plant or where it is used.

In transpiration, water only moves up the plant in one direction, from the soil into the root hair cells, and into the xylem of the roots from where it moves up to the stems, leaves and flowers. The presence of water supports these structures. In leaves the water also is used together with carbon dioxide in photosynthesis. Carbon dioxide enters the leaf by diffusion from the surrounding air.

In translocation, the initial source of sucrose is photosynthesising cells where it is formed from glucose, and the sinks are:

- respiring cells
- dividing cells
- storage cells where it is converted to starch or other storage substances.

Source-to-sink pathways alter during the seasons.

- An immature leaf is a sink of sucrose in early spring when it is growing rapidly. A mature leaf is a source in late spring and summer.
- A root is a source of sucrose in early spring (converted from starch which was stored in the cells over winter). A root is a sink in late spring and summer as it grows rapidly, and later as it starts to store starch for the winter.

Systemic pesticides are pesticides that are absorbed through the leaves of a plant and translocated to all parts of the plant through the phloem. This protects all parts of the plant from pest damage, not just the parts onto which the pesticide was sprayed. Once the pesticide has been absorbed it cannot be washed away by the rain.

Xylem

Xylem is made of a bunch of xylem vessels. Each xylem vessel comprises dead cells stacked one on top of the other with no cross walls to make a continuous tube. Xylem vessels transport water in one direction only, from root to leaf. Xylem vessels have woody cell walls that also provide support for the plant.

CAM

Water enters the roots through the root hair cells, and crosses through the root cortex (packing) cells to the xylem in the middle of the root. It travels up through the stem in the xylem and into the leaves. In the leaf it moves out of the xylem cells into the mesophyll cells, and evaporates from the surface of these cells into the air spaces.

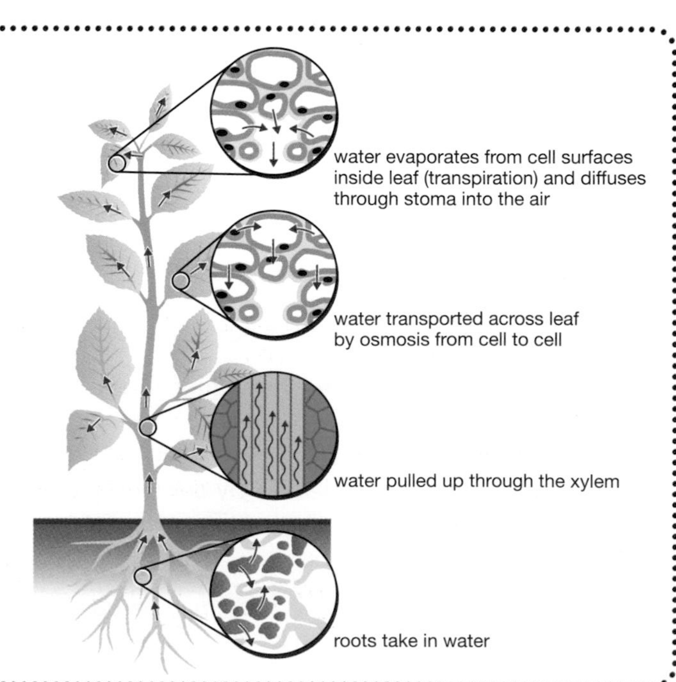

Fig. 2h.02: Transport of water through a plant

Investigating the pathway of water

The pathway of water through the above-ground parts of a plant can be investigated using a stalk of celery placed in water containing food colouring. As water moves up the plant, it carries the colouring. After a day or so, the xylem 'veins' in the leaves will show the colour.

Cutting across the stem shows that the colour is concentrated in the xylem of the stalk.

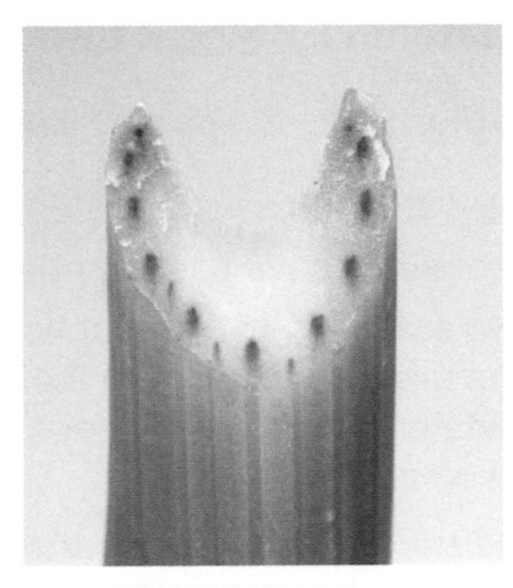

Fig. 2h.03: Food colour in the xylem of celery

Root hair cells

Root hair cells are specialised epidermal cells of plant roots and have long extensions that increase their surface area for absorption of water by osmosis and mineral ions by active transport.

The concentration of water molecules in the root cell cytoplasm is less than in the soil water, so water moves into the root hair cells by osmosis. Water molecules then move from cell to cell towards the middle of the root and into the xylem cells.

Transpiration

Transpiration is the evaporation of water from the surfaces of a plant. Transpiration occurs because water evaporates from the surfaces of the spongy mesophyll cells into the air spaces and the water vapour diffuses out through the stomata. The unidirectional movement of water up the xylem is called the transpiration stream.

Factors affecting the rate of transpiration

The rate of transpiration is affected by changes in several factors.

- **Humidity** is the concentration of water vapour in the air.
  – The lower the humidity of the air, the fewer water molecules it contains, increasing the concentration gradient for water molecules between the air spaces in the leaf and air outside. This increases the rate of diffusion of the water vapour from the stomata, hence it increases the rate of transpiration.
  – The higher the humidity of the air, the smaller the concentration gradient for water molecules between the air spaces and air. So the rate of diffusion of the water vapour from the stomata is slower, hence the rate of transpiration decreases.
- The higher the wind speed, the faster evaporating water molecules from the leaf are moved away. This maintains a high concentration gradient for water molecules between the inside of the leaf and the air, so increasing the rate of transpiration.
- At higher temperatures, molecules have more kinetic energy and move faster. This increases the rate of transpiration from the leaf. Also the higher the air temperature, the more water vapour it can hold. This increases the concentration gradient for water molecules, so transpiration rate increases as temperature increases.
- As light intensity increases, the rate of photosynthesis increases. Above a certain light intensity, the opening of the stomata is triggered. The larger the opening of the stomata, the more molecules of water vapour can leave per unit time, and so the transpiration rate increases with increasing light intensity.
- Plants can close their stomata to prevent excessive water loss and wilting. As gas exchange stops with the closing of stomata, some plants stop photosynthesising in hot, dry sunny weather if the requirement is to close the stomata to prevent water loss.

Water potential gradient through a plant

Water potential is the potential of a solution to take in water.

- Pure water has a water potential of 0.
- All solutions have a water potential less than 0 (they have negative values).
- If two solutions of different water potential are separated by a partially permeable membrane, water molecules will move from the higher water potential (where there are more water molecules) down the water potential gradient to the solution of lower water potential (where there are fewer water molecules).

Comparing negative values can be confusing, and you need to make sure you understand which solution has the higher water potential and which has the lower. For example, a solution with a water potential of −1 has a **higher** water potential than a solution with a water potential of −2.

Transpiration draws water through the plant from the roots because of the water potential gradient.

- Transpiration results in a loss of water molecules from the mesophyll cells, so their water potential falls.
- Water moves from the leaf xylem vessels to the mesophyll cells by osmosis, because the water potential of the mesophyll cells is lower than the water potential of the xylem vessels.
- Xylem vessels form a continuous tube through the plant between the leaves and the roots.
- Water molecules are cohesive, meaning they try to stick together. As water molecules in the leaf xylem move into the mesophyll cells, they 'pull' on the water molecules a little further down the xylem, pulling them up to the top.
- This pull is transmitted all the way down the xylem to the roots, where water molecules are pulled into the root xylem from the cortex cells. This produces the water potential gradient in the xylem.
- This pull reduces the water potential of the cortex cells, so water moves into them from the root hair cells by osmosis.
- This reduces the water potential of the root hair cells, so water moves into them from the soil water by osmosis.

If transpiration happens more quickly than the roots are able to supply water from the soil through the xylem to the leaves, the plant will **wilt**.

Investigating the rate of transpiration

The rate of transpiration can be investigated in different ways.

- Using a **mass potometer**: The pot and soil of a potted plant are covered with a plastic bag. The pot is then placed on a balance and the mass measured. Any loss in mass will be by transpiration. Measuring the loss of mass at different times will give the rate of transpiration.
- Using a **potometer** (see Fig. 2h.04): As water is lost by transpiration from the shoot, more water is taken up from the tube, so the level of water in the pipette falls. Measuring the level of water in the pipette at different times will give the rate of transpiration.

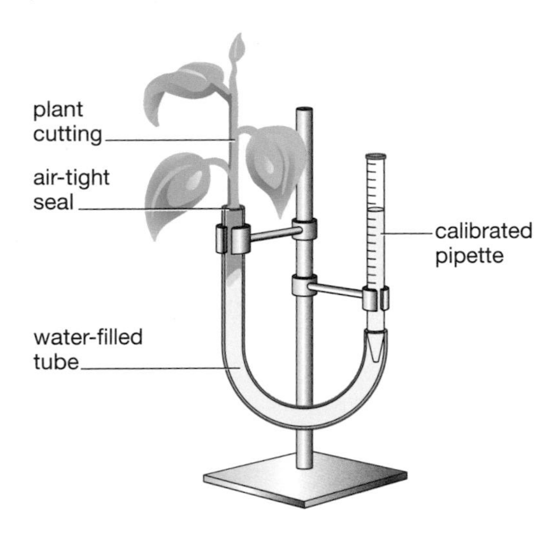

Fig. 2h.04: A potometer for measuring transpiration from a leafy shoot

The apparatus can be adjusted to show the effect of different factors:

- temperature – measure the transpiration rate at different temperatures
- wind speed – use the cold setting on a hair dryer to blow air across the shoot at different speeds
- light intensity – measure the rate of transpiration in a dark place and in the light
- humidity – place the plant in a bell jar with a beaker of water (high humidity) and measure the rate of transpiration, then place the plant in a bell jar with silica gel (or other desiccant to reduce humidity) and measure the rate of transpiration.

CAM

Adaptations of plants to environmental conditions

The root, stem and leaf sections described above are found in dicotyledonous plants that live in average conditions of temperature and water supply, such as in a UK garden.

Aquatic plants

Plants that live in a pond may float on the water, such as duckweed *(Lemna)*, or may be completely submerged with floating leaves, such as Canadian pondweed (*Elodea*). These plants need adaptations to grow well in these conditions.

small plants of one or two tiny leaves, light enough to float on water – they divide into new plants if more leaves are produced

most stomata on upper surface of leaf for gas exchange with air

simple roots absorb what the plants need from the water

Fig. 2h.05: The adaptations of duckweed

Water plants that are rooted in the bottom of a stream, lake or river usually have small root systems because the water provides most of the support to the plant. Underwater leaves usually have no protective cuticles, as they are not at risk from drying out and need to absorb nutrients directly from the water. The xylem is greatly reduced because water is absorbed directly from the surrounding water.

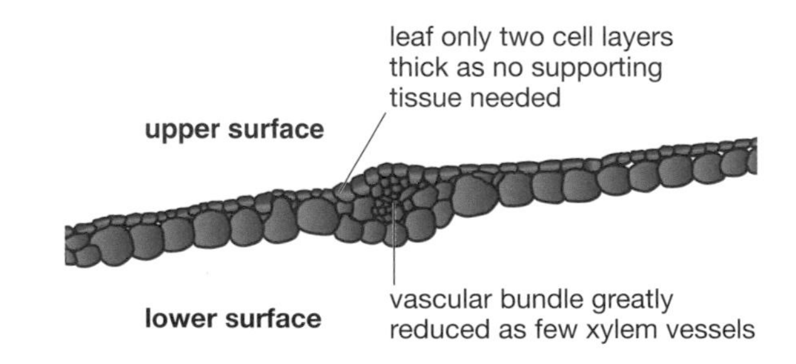

Fig. 2h.06: A section through a leaf of *Elodea* (Canadian pondweed)

Desert plants

Plants that live in the desert must be adapted to arid conditions, high temperatures during the day and often low temperatures at night.

TIP

You will be expected to apply your knowledge of adaptations in plants to local examples where possible. So focus on the environmental conditions that the example plants need to grow in, and link the adaptations of the plants to these conditions.

spines protect cactus from herbivores

rate of transpiration reduced by:
- no leaves
- stomata sunk deep into pits or ridges in stem

extensive root system to collect and store as much water as possible from the soil after rain

stem is:
- green for photosynthesis
- full of water-storage tissue (succulent) for times of drought

Fig. 2h.07: The adaptations of a cactus to arid conditions and high temperatures

Composition of human blood

Blood is a liquid tissue that includes different cell types suspended in **plasma**. Blood cells enclose solutes that may affect the properties of the plasma; therefore chemical environments inside the cells can be different from the plasma.

Blood plasma

Blood plasma is responsible for the transport of heat energy and many soluble substances around the body, including:

- carbon dioxide – diffused out of aerobically respiring cells into the plasma for transport to the lungs where it is excreted
- soluble digested food molecules, such as glucose, amino acids, glycerol and fatty acids – absorbed from the small intestine, for transport to cells that need them
- urea – formed in the liver from the breakdown of excess amino acids, transported to the kidneys for excretion
- hormones – secreted by glands directly into the blood plasma and transported to their target organs
- AM plasma proteins – carry some substances that are not soluble in plasma, such as fatty acids and steroid hormones such as testosterone and oestrogen; they also help maintain the blood's water potential
- antibodies – made by white blood cells to fight infection
- heat energy – released by respiration from cells in deeper organs and transported around the body and to the surface where it is transferred to the environment.

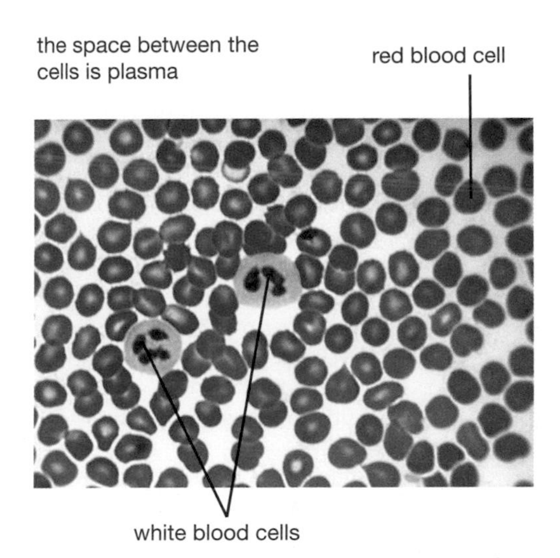

Fig. 2h.08: A blood smear showing white and red blood cells

Adaptations of red blood cells

Red blood cells carry oxygen around the body, attached to molecules of **haemoglobin i**n the cells. Red blood cells have many adaptations for this function.

- Red blood cells have a disc-like shape that is thinner in the middle than at the edge. This increases the surface area over which oxygen can diffuse into and out of the cell. This shape means red blood cells can bend and squeeze through capillaries.
- The structure of the cell membrane of red blood cells makes them more resistant than other cells to bending and to friction, to protect them as they squeeze through narrow capillaries.
- Red blood cells have no nucleus, so there is more room for haemoglobin in the cell.
- Haemoglobin is a molecule that bonds with oxygen when the oxygen concentration is high (capillaries in the lungs) and gives up oxygen when the oxygen concentration is low (capillaries in tissues). One red blood cell contains about 250 million molecules of haemoglobin, and each haemoglobin molecule carries four molecules of oxygen.

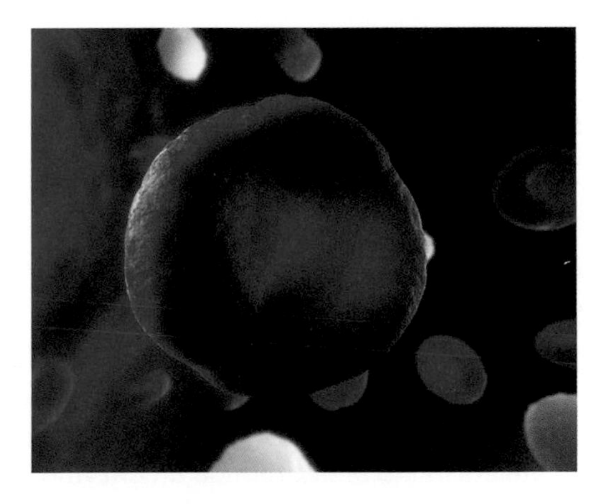

Fig. 2h.09: A red blood cell

White blood cells and the immune system

The **immune system** helps to protect us from infection by attacking pathogens that enter the body. Two types of white blood cells are important in this system.

- **Phagocytes** are white blood cells that **engulf** (flow around and 'swallow') and kill pathogens.
- **Lymphocytes** make soluble proteins called **antibodies**, which they release into the blood plasma. There are millions of different types of lymphocytes, and each one is able to make a specific antibody. Antibodies can make the cell membranes of the pathogens burst, or they can prepare pathogens for phagocytosis, e.g. by making the pathogens clump together. Lymphocytes also make antitoxins which counteract the toxins released by some pathogens.

CAM

The immune system attacks any cells in the body that are 'foreign' and not of the body ('self'). This is helpful in preventing infection, but not helpful in tissue transplants when an organ or tissue from a different body is placed in a patient (usually to replace damaged tissue). The immune system responds by attacking and killing the new tissue, causing **tissue rejection**. To avoid tissue rejection, the immune system needs to be suppressed with drugs.

How vaccination makes you immune

A **vaccination** is when a non-infective form of a pathogen is given to a person, by injection or by mouth. This stimulates the immune system to respond to the pathogen without making the person ill.

Therefore a vaccination makes you **immune** to a disease without having to have it first.

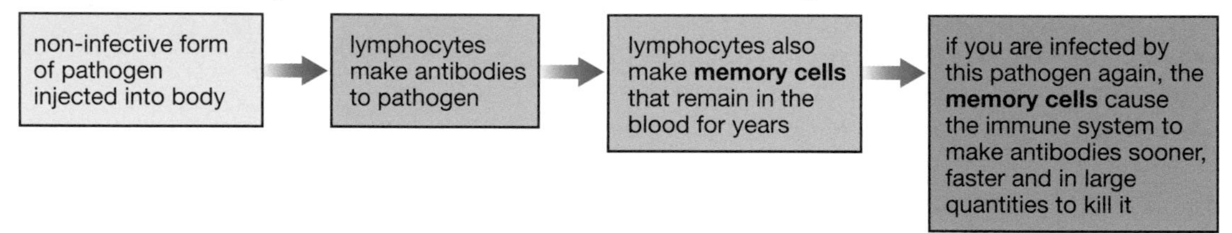

Fig. 2h.10: The effect of vaccination

Platelets and blood clotting

Platelets are tiny pieces of cell that respond to injured blood vessels by causing the blood to clot. This seals the damage and prevents:

- further blood loss
- the entry of pathogens, so reducing the risk of infection.

CAM

Liver cells produce a soluble protein called fibrinogen that is secreted into the blood plasma. When a blood vessel is damaged, platelets in the area produce an enzyme that breaks down the fibrinogen to another protein called **fibrin**. This protein is insoluble and forms long fibres that make a mesh in which blood cells are trapped, and so forms a blood clot.

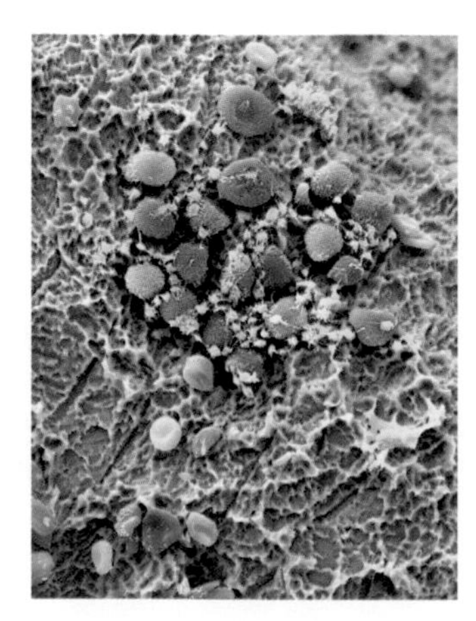

Fig. 2h.11: A false colour scanning electron microscope image of a blood clot; the platelets are blue and red blood cells are red

Structure and function of the heart

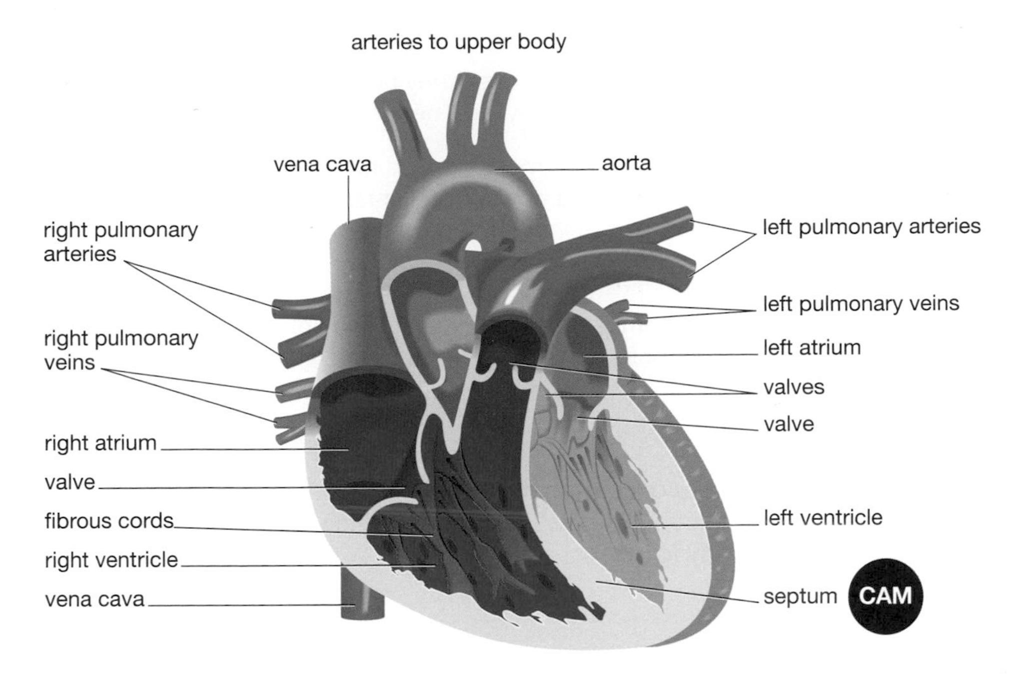

Fig. 2h.12: The human heart

The heart is effectively two separate pumps, one on the right and one on the left.

- The right side receives deoxygenated blood from the body through the **vena cava** and pumps it to the lungs through the pulmonary artery.
- The left side receives oxygenated blood from the lungs through the pulmonary vein and pumps it to the rest of the body through the **aorta**.

The structure of the heart supports its function.

- The heart walls are muscles, for contraction.
- The two **atria** collect blood entering the heart, contract and pass the blood to the ventricles.
- The two thick muscular walls of the **ventricles** contract and force blood out of the heart through arteries.
- The muscle of the left ventricle is thicker and more powerful than that of the right ventricle as it needs to pump the blood over a greater distance.
- To make sure that the blood always flows in one direction, the heart has **valves** between the atria and ventricles and at the base of the arteries – the valves close to prevent backflow, but open when the blood is being pumped in the right direction.
- CAM The septum separates the two ventricles so blood on each side does not mix.

Changing heart rate

The effect of exercise

During exercise, muscle cells are respiring faster because they need energy more rapidly for contraction. Therefore they need oxygen and glucose more rapidly, and need to get rid of carbon dioxide more rapidly.

The blood carries oxygen and glucose to respiring muscles, and takes carbon dioxide away. During exercise, the heart beats faster to circulate the blood to and from the muscles faster. The heart rate increases during exercise.

The effect of adrenaline

Adrenaline is a hormone secreted during times of fright, stress or the anticipation of action. It increases the heart rate, preparing the body for a quick response. (Therefore, it is called the 'fight or flight' hormone.)

Adrenaline also increases the force of contraction of the heart, and causes an increase in blood sugar concentration by stimulating the conversion of glycogen to glucose in liver and muscle cells. These changes help to prepare the body for action.

Blood vessels

Structure and function

The blood vessels of the body can be divided into three types: **arteries**, **veins** and **capillaries**.

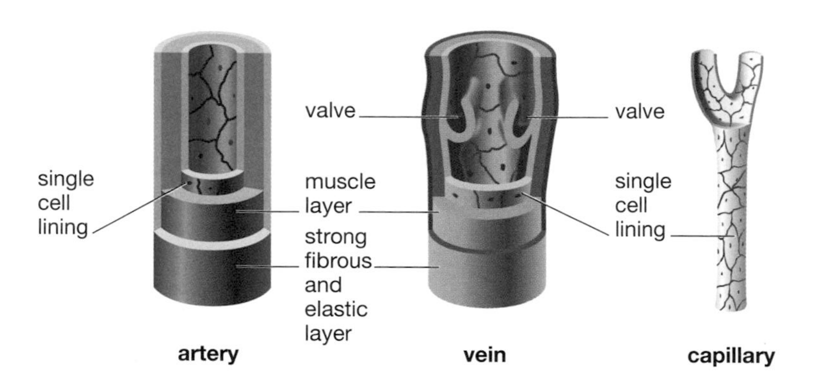

Fig. 2h.13: Artery, vein and capillary; the muscle and fibrous layers are thicker in an artery than in a vein

	Structure	Function
Arteries	• thick walls, with a thick elastic layer on the outside and thick muscle layer inside • small lumen (centre where blood flows)	• carry blood *away from* the heart • elastic tissue in walls is stretched as blood is pumped into them and recoils when heart relaxes, keeping blood pressure high • tough outer layer prevents artery bursting
Veins	• thinner walls than arteries • little muscle in wall • large lumen • valves on inner walls	• carry blood *back towards* the heart • large lumen so blood flows easily • valves prevent backflow of blood, so blood only flows towards heart
Capillaries	• walls single cell layer thick	• where substances are exchanged with tissues • thin cells increase the rate of exchange by keeping distance to a minimum

The circulatory system

The circulatory system consists of the heart (the pump) and the continuous blood vessels that carry the blood around the body and through tissues.

The arteries and veins of the human circulatory system are named according to the organ they supply:

- lungs – pulmonary arteries and veins
- liver – hepatic arteries and veins
- kidneys – renal arteries and veins
- heart – coronary arteries and veins direct to the heart muscle.

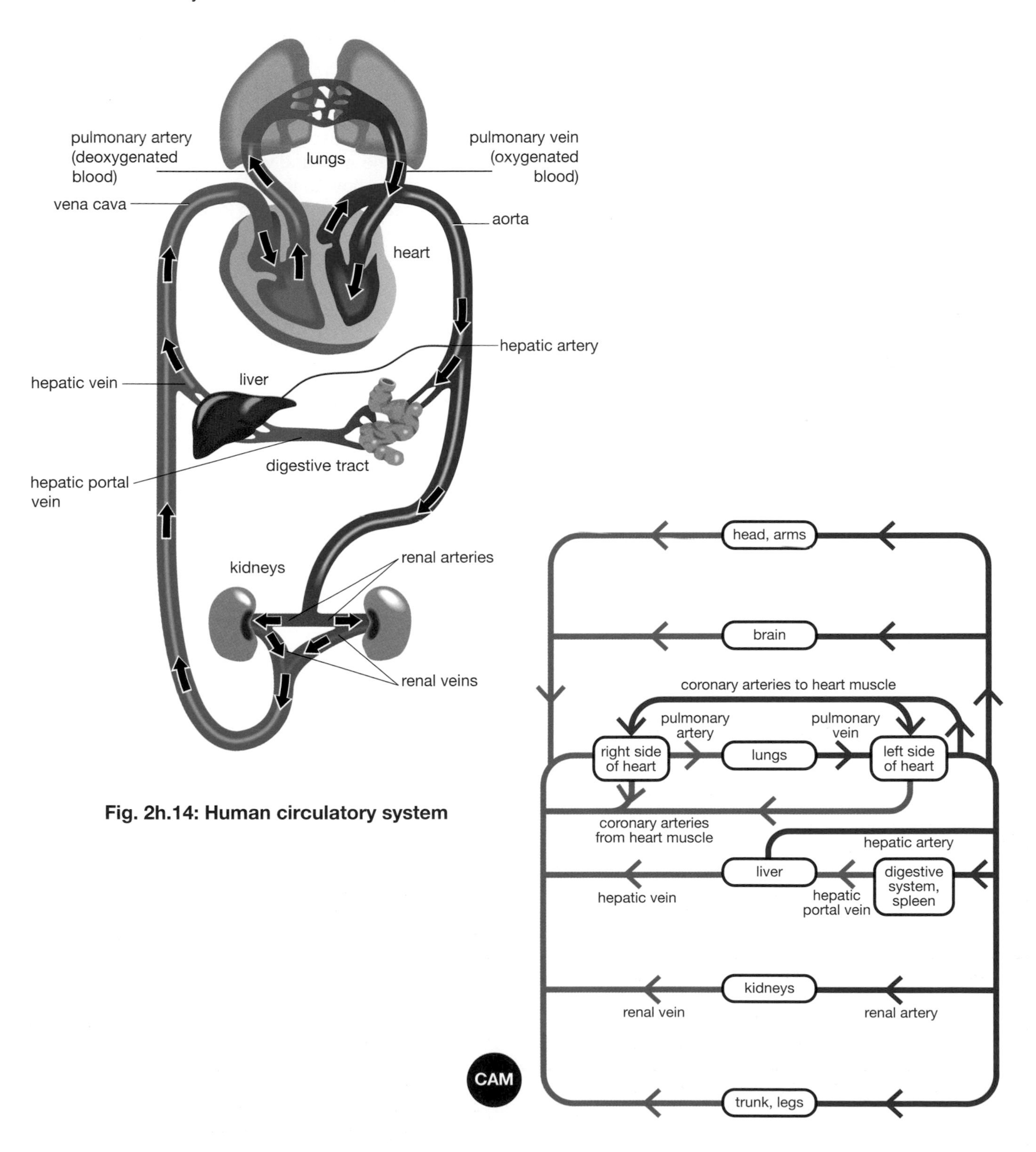

Fig. 2h.14: Human circulatory system

Note that the only blood vessel that connects two organs directly is the **hepatic portal vein** which carries digested food molecules from the intestine straight to the liver.

A double circulatory system

The human circulatory system is a **double circulation**, with one circulation through the lungs and a separate circulation through the body. This is an advantage because:

- blood flowing through the lung circulation can be at a lower pressure, so preventing damage to the delicate capillaries in the lungs
- blood flowing through the body circulation can start at a much higher pressure, making it easier for it to be pumped to the far ends of the body and back.

Coronary heart disease

The coronary arteries supply the heart muscle cells with oxygen and glucose for respiration (because the muscle cells cannot absorb these from the blood pumping through the atria and ventricles). If cholesterol is deposited on the inside of these arteries, this may reduce the blood flow to the muscle cells and cause heart pains. If a coronary artery becomes blocked, heart muscle cells may die due to lack of oxygen, causing a heart attack, which is a form of **coronary heart disease**.

Blockage of coronary arteries may occur as a result of:

- a diet high in saturated fats (fats from animal sources)
- smoking
- long-term stress
- decrease in the amount of exercise.

The lymphatic system

Fluid leaving the capillaries in tissues becomes tissue fluid. Cells exchange substances directly with the tissue fluid by diffusion, osmosis and active transport.

Some of the substances transferred to the tissue fluid are taken up by the capillaries. The rest is taken up by another set of tubes called the **lymphatic system**. The fluid inside this system is called **lymph**.

The lymphatic system carries the lymph back through the body to near the heart, where it is returned to the blood.

Tissues associated with the lymphatic system, such as in bone marrow, make lymphocytes. Lymphocytes are transferred to the blood via the lymph.

You should now be able to:

- ★ explain why simple unicellular organisms can rely on diffusion for movement of substances in and out of the cell but multicellular organisms need a transport system (see page 71)
- ★ describe the role of phloem in transporting sucrose and amino acids in a plant (see page 71)
- ★ describe the role of xylem in transporting water and mineral salts in a plant (see page 72)
- CAM ★ sketch a diagram to show the positions of xylem and phloem tissues in a root, stem and leaf (see page 71)
- ★ explain how water is absorbed by root hair cells (see page 73)
- CAM ★ state the pathway taken by water through the root, stem and leaf (see page 72)
- ★ define transpiration and explain how it occurs (see page 73)
- CAM ★ explain the role of a water potential gradient and cohesion of water molecules in transpiration (see page 74)
- ★ explain how the rate of transpiration is affected by changes in humidity, wind speed, temperature and light intensity (see page 75)
- CAM ★ define translocation in terms of the movement from sources to sinks (see page 72)
- ★ describe the translocation throughout the plant of systemic pesticides (see page 72)
- ★ describe and explain the adaptations of plants that live in these environments: a pond, a garden, a desert (see page 76)
- ★ state the composition of blood, and describe the functions of each part (see page 77)
- CAM ★ describe how the immune system responds to disease and causes tissue rejection (see page 78)
- ★ describe how vaccination enables the body to respond to a pathogen more effectively (see page 78)
- ★ describe the structure and functions of different parts of the heart (see page 79)
- ★ describe and explain the effect of exercise and adrenaline on heart rate (see page 80)
- ★ describe the structure of arteries, veins and capillaries and explain how structure is related to function (see page 80)
- ★ name the blood vessels to and from the heart, the lungs, the liver and the kidneys (see page 81)
- CAM ★ explain the advantage of a double circulation system in humans (see page 82).
- ★ describe the effect of blockage of coronary arteries and state some possible causes (see page 82)
- ★ describe the function of the lymphatic system (see page 82)
- ★ describe the transfer of substances between capillaries and tissue fluid (see page 82).

Practice questions

1.

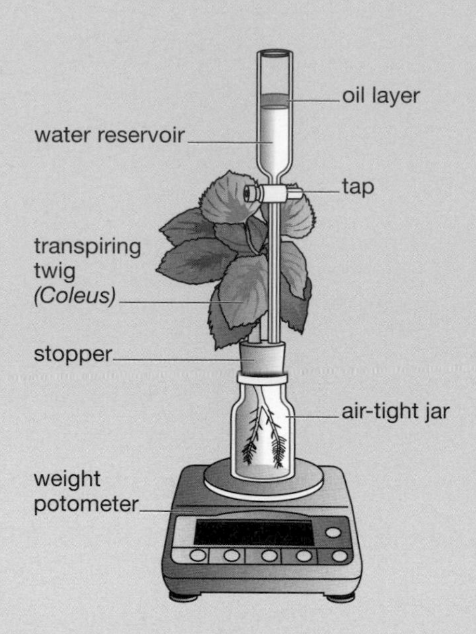

A plant was set up (as illustrated) for an investigation on the rate of transpiration over a 24-hour period.

Table of results:

	Mass of apparatus (g)	**Volume of water in measuring cylinder (cm^3)**
Start	235	101
Finish	207	72

(a) Calculate the loss of mass during the 24-hour period. **(1)**

(b) Calculate the volume of water absorbed by the plant during the 24-hour period. What **mass** of water was taken up by roots during this time? **(2)**

(c) Referring to how water moves up a plant, explain why your answers to (a) and (b) are so similar. **(4)**

(d) Give one way in which the plant uses water it absorbs through its roots. **(1)**

(e) Explain why this plant may lose a different amount of water on another day. **(2)**

(f) What would happen to the plant if it lost more water than it absorbed? **(2)**

2. (a) The heart acts as a double pump. Describe the structure and the function of the right hand side of the heart. **(5)**

CAM

(b) Describe the causes of a coronary heart attack. **(2)**

(c) How might smoking have caused a heart attack? **(3)**

(d) How might diet have caused a heart attack? **(2)**

(e) Describe how oxygen absorbed in your lungs reaches your respiring cells. **(5)**

I Excretion

You will be expected to:

- ★ state that, in flowering plants, carbon dioxide is a waste product of respiration and oxygen is a waste product of photosynthesis, and that these gases are lost to the air through leaf stomata
- ★ state that, in humans, the lungs, kidneys and skin are organs of excretion
- ★ describe the functions of the kidneys as excretion and osmoregulation
- ★ describe the structure of the urinary system
- ★ describe the structure of a nephron
- ★ describe ultrafiltration
- ★ describe the reabsorption of water and glucose from the nephron
- ★ describe the role of ADH in regulating blood water content
- ★ name the main constituents of urine

- ★ describe the role of the liver in the formation of urea and breakdown of alcohol, drugs and hormones
- ★ explain dialysis
- ★ discuss the advantages and disadvantages of a kidney transplant.

Excretion in flowering plants

Waste products of metabolism need to be **excreted** from an organism, to prevent them affecting other cell processes. In a plant, the waste products are:

- carbon dioxide from respiration
- oxygen from photosynthesis.

The organ of excretion in plants is the leaf, as these gases leave the plant through the stomata.

Excretion in humans

In a human, the organs of excretion are:

- lungs – excrete carbon dioxide from respiration
- skin – secretes sweat to cool down, however, sweat may contain salts that are excreted
- kidneys – excrete **urea** produced in the liver from excess amino acids, also mineral salts, in a solution called **urine**.

The kidneys also excrete other substances that have been broken down in the liver. These include the breakdown products of alcohol, drugs and some hormones. This protects the body from unwanted effects of these substances.

The kidneys also carry out **osmoregulation**, the regulation of water content in the body, by removing excess water from the blood. This prevents too much water entering cells and bursting them, or too much water leaving cells causing them to shrink, affecting cell processes.

Human urinary system

The human urinary system includes the following organs.

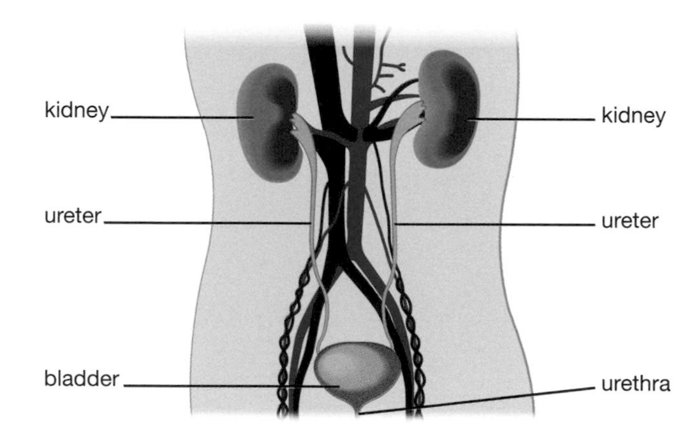

Fig. 2i.01: The human urinary system

Organ	Structure	Function
kidneys	two bean-shaped organs at the back of the abdominal cavity, on either side of the spine	carry out excretion and osmoregulation to form urine
ureters	tube from kidney to bladder	carry urine to bladder
bladder	muscular sac	stores urine temporarily until expelled from body
urethra	muscular tube from bladder to outside world	through which urine is expelled from body

Nephron structure

Each kidney is full of capillaries and about 1.3 million nephrons (kidney tubules). Nephrons are where ultrafiltration and reabsorption take place to form urine.

The nephron is closely associated with a blood capillary, with which substances are continually exchanged.

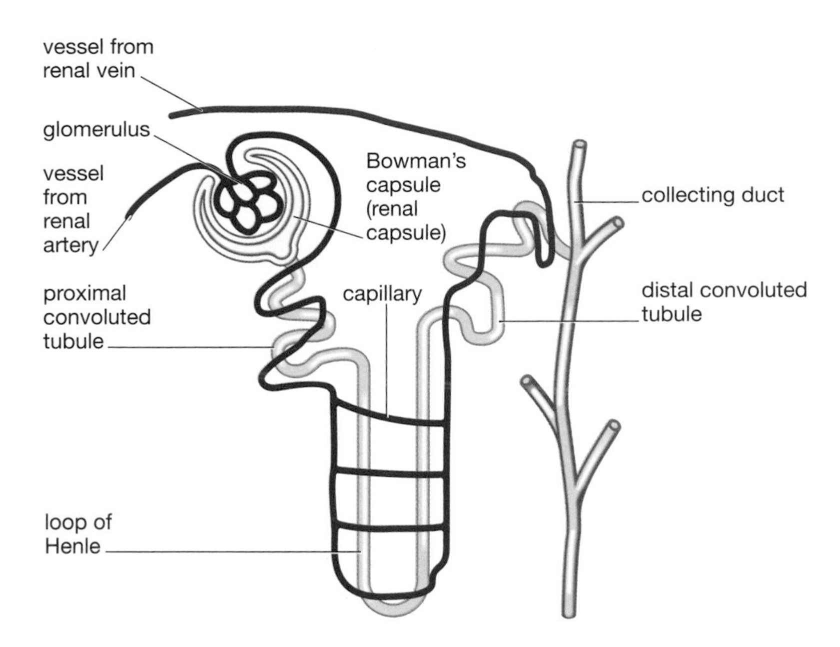

Fig. 2i.02: Structure of a nephron

Different activities take place in different parts of the nephron:

- ultrafiltration (filtration) in the Bowman's (renal) capsule
- selective reabsorption of glucose (and other products of digestion) in the proximal convoluted tubule
- selective reabsorption of dissolved ions needed by the body in all parts of the nephron
- reabsorption of water from the collecting duct.

CAM

Each kidney has an outer region called the **cortex** and an inner region called the **medulla**. The renal capsule of a tubule is situated in the cortex; the loop of Henle dips down into the medulla and returns to the cortex; the collecting duct passes back through the medulla to the centre of the kidney.

Ultrafiltration

The Bowman's capsule is a cup-shaped structure at the start of each nephron. A knot of special 'leaky' capillaries called a **glomerulus** sits inside the Bowman's capsule.

Fluid containing water and dissolved solutes from the blood plasma is forced out of the glomerulus into the capsule. Large protein molecules and blood cells are too large to get out of the capillaries.

This process is called **ultrafiltration** because it is filtration on the scale of molecules. The fluid that enters the nephron is called the **glomerular filtrate**.

The glomerular filtrate is very similar to the blood plasma: it contains water, urea, salts, amino acids, fatty acids, glycerol and glucose.

Selective reabsorption of glucose

Glucose is essential to the body for respiration. In a healthy kidney, all glucose is reabsorbed from the tubule filtrate into the blood capillary in the region of the proximal convoluted tubule. This process involves active transport, and uses energy from respiration.

This is **selective reabsorption** because glucose is being selected from the filtrate for reabsorption.

Other soluble products of digestion are also reabsorbed: amino acids, fatty acids and glycerol.

Water reabsorption and ADH

Water is reabsorbed from the tubule across the wall of the collecting duct. The reabsorbed water moves into the blood in the capillaries.

How much water is reabsorbed depends on the water concentration of the blood, and is controlled by the hormone **ADH** (antidiuretic hormone).

If blood water content is too *low* (e.g. when you become dehydrated or when you eat salty foods):

- more ADH is secreted by the pituitary gland at the base of the brain into the blood and carried to its target organs, the kidneys
- ADH causes the walls of the collecting ducts of the nephrons to become more permeable to water
- more water moves out of the collecting ducts back into the blood in the capillaries
- the result is a smaller volume of concentrated urine.

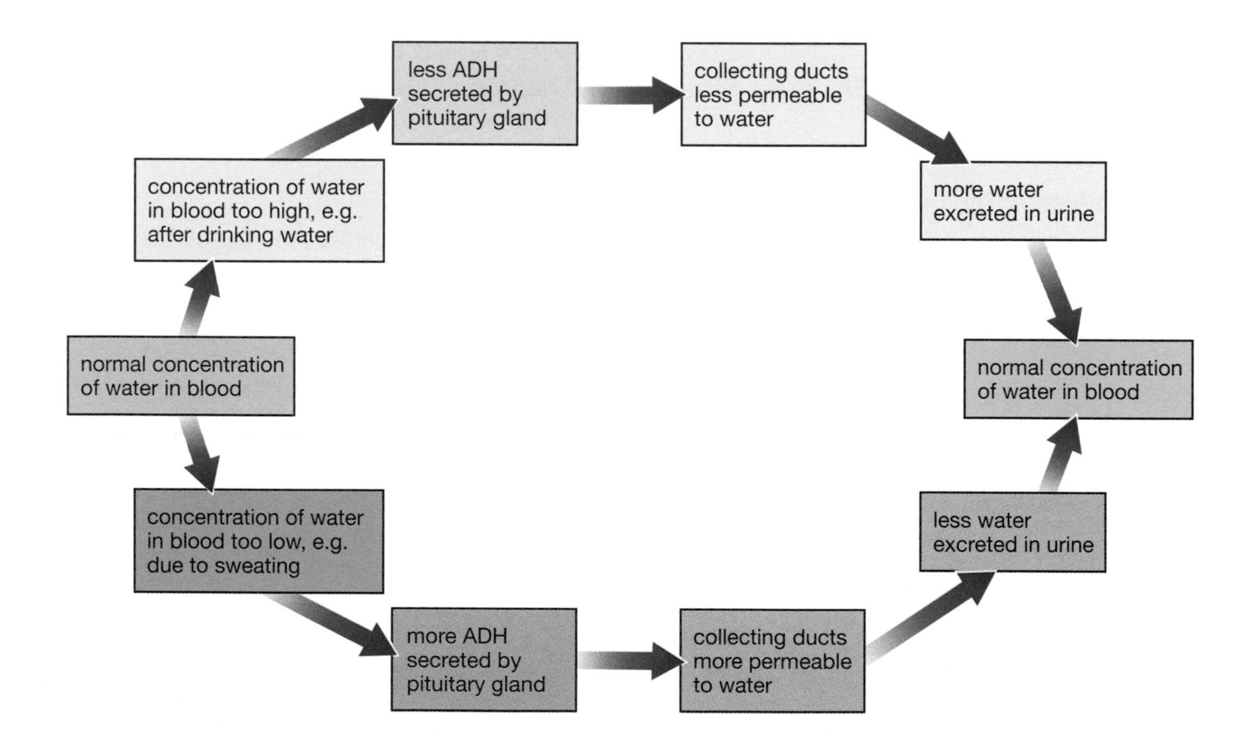

Fig. 2i.03: Blood water concentration control

If the blood water content is too *high* (e.g. when you drink a large quantity of water):

- less ADH is secreted by the pituitary gland into the blood
- the collecting ducts remain comparatively impermeable to water
- a smaller amount of water moves out of the collecting ducts back into the blood
- the result is a larger volume of more dilute urine.

Treating kidney failure

Kidney failure occurs when both kidneys are unable to carry out the processes of filtration and reabsorption properly. (You only need one kidney most of the time.) This is dangerous because waste products and excess substances rapidly build up in the blood, which can damage cells.

Kidney failure can be treated by:

- **dialysis**: linking an artery to a dialysis machine and passing blood through a solution before returning it to the body through a vein
- **transplant**: attaching a healthy kidney to a blood vessel in the patient's body to carry out filtration and reabsorption naturally.

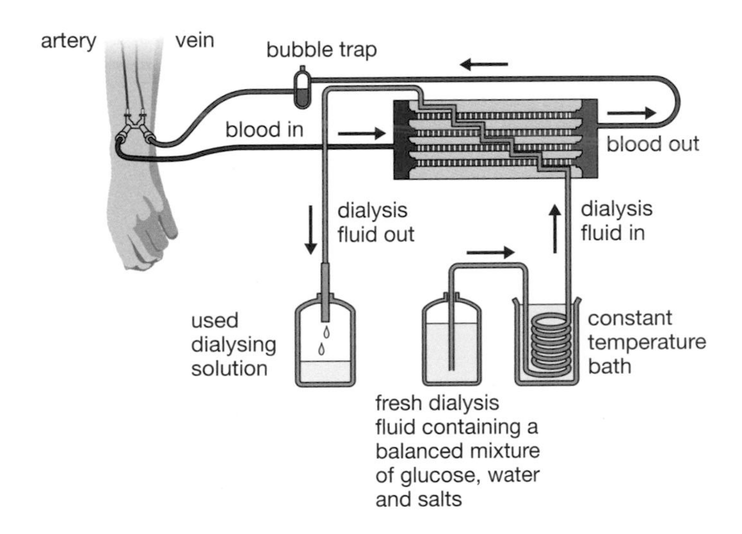

Fig. 2i.04: Haemodialysis – substances such as water, urea and mineral salts are exchanged between the blood and dialysis fluid to restore their correct balance in the blood

Each of these treatments has advantages and disadvantages.

Treatment	Advantages	Disadvantages
dialysis	• readily available in hospitals • reduces risk of damage to cells by high concentrations of waste products and excess substances	• takes several hours and needs to be done several times a week • risk of infection where tube is attached to arm
kidney transplant	• reduces risk of damage to cells by high concentrations of waste products and excess substances • more natural control of blood concentrations, so safer	• can only be done if new kidney is similar in tissue type to patient's body tissues, so may take a long time to get a good match • immune system must be suppressed with drugs to prevent rejection of kidney, increasing number of infections by pathogens

You should now be able to:

★ state the waste metabolic products of plants and the organ of their excretion (see page 85)

★ state the organs of excretion in humans (see page 85)

★ describe how and where urea is formed in the body (see page 86)

★ state some substances found in urine (see page 85)

★ name the organs of the urinary system (see page 86)

★ name the two roles of kidneys (see page 86)

★ describe the structure of a nephron (see page 87)

★ define the term *selective reabsorption* with reference to where glucose is reabsorbed in a nephron (see page 88)

★ describe where water is reabsorbed in a nephron, and describe the role of ADH in regulating the water content of the blood (see page 88)

CAM ★ explain how dialysis can maintain the correct concentration of substances in the blood (see page 89)

★ identify the advantages and disadvantages of kidney transplants, compared with dialysis (see page 89).

Practice questions

1. Below is a diagram of the kidney tubule.

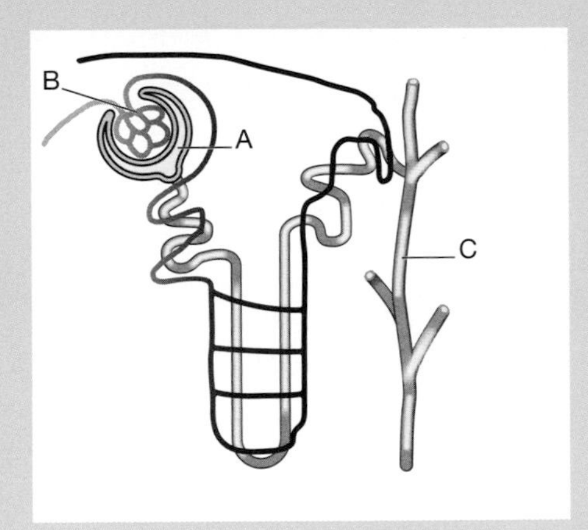

(a) Label A, B and C. **(3)**

(b) Name five substances filtered out of the blood and which enter B. **(5)**

(c) What two features of these substances make it possible for them to be filtered out of the blood? **(2)**

(d) List four substances that are reabsorbed back into the blood from the tubule. **(4)**

(e) What three substances are in urine? **(3)**

(f) Where is urine stored before urination? **(1)**

CAM **2.** Sometimes kidneys fail because of accident or disease. A person with kidney failure can be treated by dialysis or have a kidney transplant.

(a) During dialysis, a person can eat as much salt, and drink as much water, as they want. Why? **(2)**

(b) Give two advantages of using a kidney transplant rather than a dialysis machine. **(2)**

3. (a) Explain why the concentration of urea in the bladder is higher than that filtered from the blood in the kidney tubules. **(1)**

(b) What type of substance is ADH? **(1)**

(c) What gland produces ADH? **(1)**

(d) What stimulus is necessary to cause ADH to be secreted? **(1)**

(e) Describe the effect of a decrease in ADH production on the kidney and on the composition of the urine. **(3)**

J Coordination and response

You will be expected to:

- ★ give examples of how organisms respond to the environment
- CAM ★ define sense organs as groups of receptor cells that respond to stimuli such as light, sound, touch
- ★ define the term *homeostasis*
- ★ describe the homeostatic control of body water content and body temperature
- CAM ★ explain control by negative feedback
- ★ describe how a stimulus, receptor and effector can result in a coordinated response
- ★ describe responses of plants to the environment including geotropic and phototropic responses
- CAM ★ explain the use of auxin in weedkillers
- ★ identify nervous and hormonal communication in humans and compare the systems
- ★ describe the structure of the human nervous system and explain how it responds to stimuli
- CAM ★ identify different types of neurones in diagrams
- ★ describe the action of antagonistic muscles
- ★ describe a simple reflex arc
- CAM ★ distinguish between voluntary and involuntary actions
- ★ describe the structure and function of the eye
- CAM ★ distinguish between the function and distribution of rods and cones in the eye
- ★ explain how the eye focuses on near and far objects
- ★ explain how the eye responds to changing light intensity
- ★ describe the role of skin in temperature regulation
- CAM ★ identify structures in skin that are associated with body temperature regulation
- ★ define the term *hormone*
- ★ describe the sources, roles and effects of some hormones
- CAM ★ discuss the use of hormones in food production
- ★ define the term *drug*
- ★ describe the use of antibiotics, and explain why they do not kill viruses
- ★ describe the effects and abuse of heroin
- ★ describe the effects of excessive consumption of alcohol
- ★ describe the toxic effects of smoking.

Responding to changes in the environment

The environment surrounding an organism changes. Some of these changes are **stimuli** (sing.: stimulus), e.g. light intensity, temperature, sound, which are detected by **sense organs** in an organism. These external or internal changes result in a **response**: a muscle contraction/relaxation or secretion of hormones in animals (in plants, the secretion of auxin that causes a differential rate of cell division). The response is triggered by an **effector**: muscles that produce movement or glands that produce hormones (in plants, any hormones that result in changes in growth).

CAM

Human sense organ	Stimulus
eye	light: colour (cone cells), intensity (rod cells)
ear	sound
pressure receptors in skin	touch
heat receptors in skin	temperature
taste buds on tongue	chemicals in food and drink
nose	chemicals in air

Homeostasis

Homeostasis means keeping the internal environment as nearly constant as possible, so that cells can operate at maximum efficiency. This requires monitoring of conditions in the body, detecting and correcting changes so that conditions are returned to the normal level.

Homeostasis of body water content

In humans, water content of the blood is constantly monitored by the pituitary gland in the brain. Water content is maintained by increasing or decreasing the secretion of the hormone ADH (antidiuretic hormone) from the pituitary gland. ADH controls the amount of water excreted in urine by affecting the permeability of the collecting duct of nephrons (see page 88).

Homeostasis of core body temperature

The skin has a major role in controlling body temperature, involving the hairs, sweat glands, blood vessels and temperature receptors. Fat below the skin can act as an insulator.

Fig. 2j.01: Structures in the human skin involved in temperature regulation

Temperature receptors in the brain detect the temperature of blood flowing past them. The brain also receives nerve impulses from temperature receptors in the skin that provide information about external temperature.

If blood temperature is too high, the brain sends impulses that cause:

- the skin's sweat glands to secrete sweat onto the skin surface – the evaporation of sweat takes heat energy from the skin surface, cooling it down
- **vasodilation** of blood vessels near the skin surface so they carry more blood – this increases the rate of heat loss by conduction, convection and radiation from the skin.

If blood temperature is too low the brain sends impulses that cause:

- **vasoconstriction** of blood vessels near the skin surface so more blood passes through blood vessels deeper in the skin – this reduces the rate of heat loss from the skin surface
- sweating to stop
- body hairs to rise – in humans, this does little more than cause goosebumps, but it provides extra insulation in furry mammals and feathered birds
- **shivering**, where muscles contract uncontrollably, increasing the rate of respiration and therefore the release of heat energy.

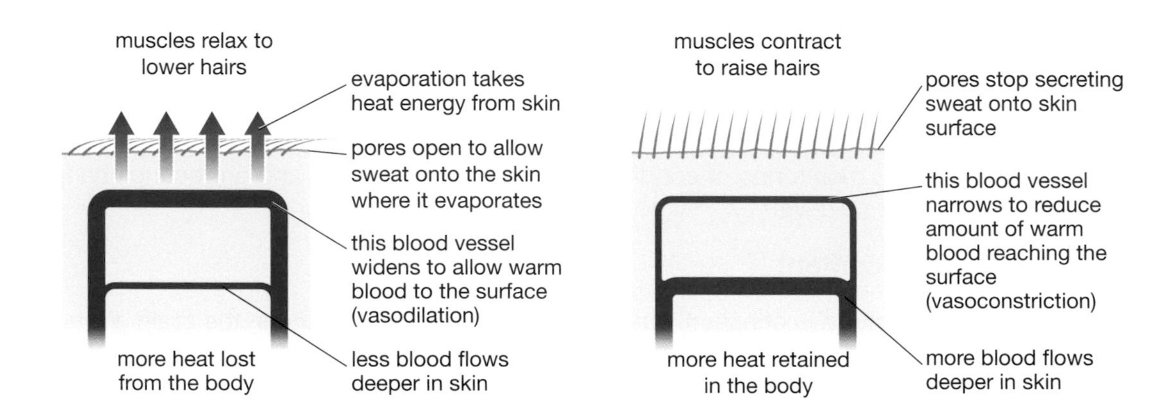

Fig. 2j.02: Changes in the skin when the blood temperature is (Left) too high (Right) too low

Responses in flowering plants

Plants respond to stimuli in the environment. For example, stomata in leaves usually close at night when there is not sufficient light for photosynthesis.

Plants also change how they grow in response to stimuli from the external environment, e.g. light intensity and direction, gravity, sources of water. A growth response in a plant to a stimulus is called a **tropism**.

Geotropism

Geotropism is the growth of a plant in response to gravity. If a seedling is turned upside down, the stem will begin to grow upwards and the root downwards.

- Stems are negatively geotropic.
- Roots are positively geotropic.

Phototropism

Phototropism is the growth of a plant in response to light. If seedlings are placed in one-sided light, the stems will grow towards the light and the roots will grow away from the light.

- Stems are positively phototropic.
- Roots are negatively phototropic.

These changes in growth help the plant shoot to get more light, and the plant root to get more water and mineral ions from the soil. These tropisms increase the chance that the plant will survive.

Auxin

Auxin is a plant growth hormone that is made in the tips of plant shoots and roots. It diffuses back along the root or shoot and affects the cells. In shoots, it stimulates cells to grow longer; in roots, it reduces how long the cells become.

In a shoot that is growing in one-sided light:

- light on one side causes the auxin to collect on the darker side of the shoot
- so the cells on the darker side grow faster than those on the light side
- the differential growth of cells on each side of the shoot causes it to grow towards the light.

In a root that is growing horizontally:

- gravity causes the auxin to collect on the lower side of the root
- so cells on the lower side grow more slowly than the ones on the upper side
- the differential growth of the root cells causes the root to grow downwards.

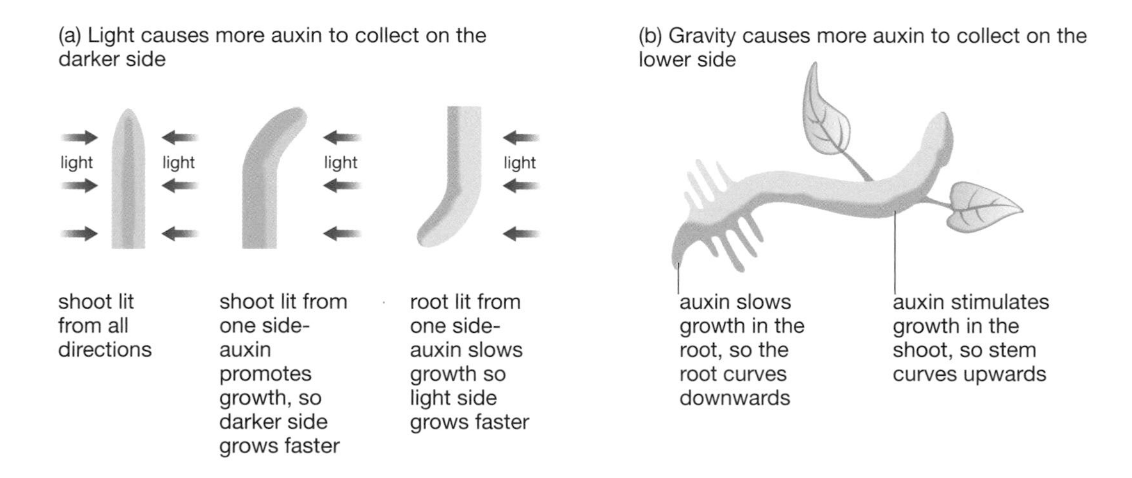

Fig. 2j.03: The effect of auxin on growth in a shoot and root (Left) in light from the side (Right) by gravity

Using auxins

Auxins are used in some synthetic **weedkillers** (herbicides). When sprayed onto plants, the auxins are absorbed by the leaves of broad-leaved plants (e.g. weed plants in a crop) but not as much by grass-like plants with narrow leaves (e.g. crops of wheat, barley or maize) which have a smaller leaf surface area.

The large amounts of auxin absorbed by the weeds cause them to grow so quickly that they cannot photosynthesise fast enough or absorb enough nutrients from the soil. So the weeds die, but the crop lives.

Human nervous system and hormonal communication

The human nervous system consists of:

- the **central nervous system** (brain and spinal cord) that controls and coordinates responses
- nerves that link the central nervous system to sensor organs and effectors (called the **peripheral nervous system**).

The nervous system produces rapid responses to stimuli, by passing electrical impulses from a sensor organ to the central nervous system, which passes electrical impulses to one or more effectors to cause the response. Nerve impulses are transmitted across a synapse by chemicals.

The human hormonal system consists of:

- endocrine glands that respond to a stimulus by secreting chemicals called **hormones** into the blood. The blood carries the hormones to **target organs** where they cause a response.

The hormonal system reacts more slowly to a stimulus than the nervous system, using chemicals to produce a response often in more than one target organ in different parts of the body. The response may also last for much longer than a response from the nervous system. The response will last until the hormone is broken down by the liver.

Neurones

The nervous system is formed from specialised cells called **neurones**, which carry electrical impulses quickly in one direction only. They have fine extensions at each end to connect with other neurones and transmit impulses to them. There are three main types of neurones.

Type of neurone	Function	Adaptations
sensory	connects receptors in sense organ to central nervous system	long extensions called axons, that group together to form a nerve
relay	only in central nervous system, links sensory to motor neurones or other relay neurones	many extensions to link to many other neurones for coordination of response
motor	connects central nervous system to effector – muscle or gland	long extensions, called axons, that group together to form a nerve

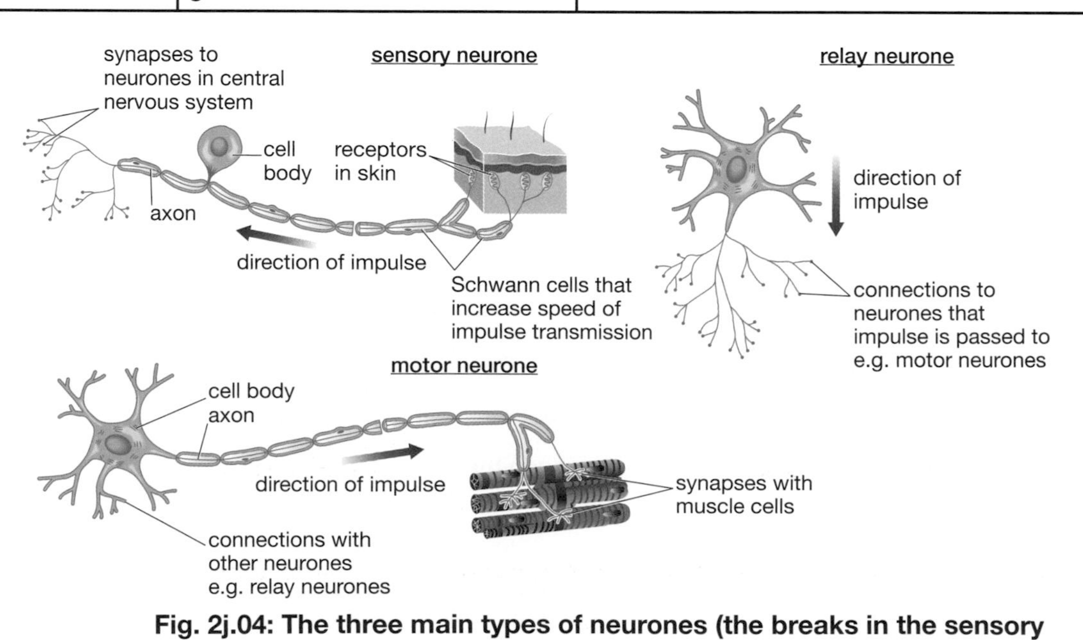

Fig. 2j.04: The three main types of neurones (the breaks in the sensory and motor neurones indicate they are much longer than shown)

Reflex arc

A **reflex arc** is the simplest level of nervous coordination in humans, involving only three neurones and three synapses.

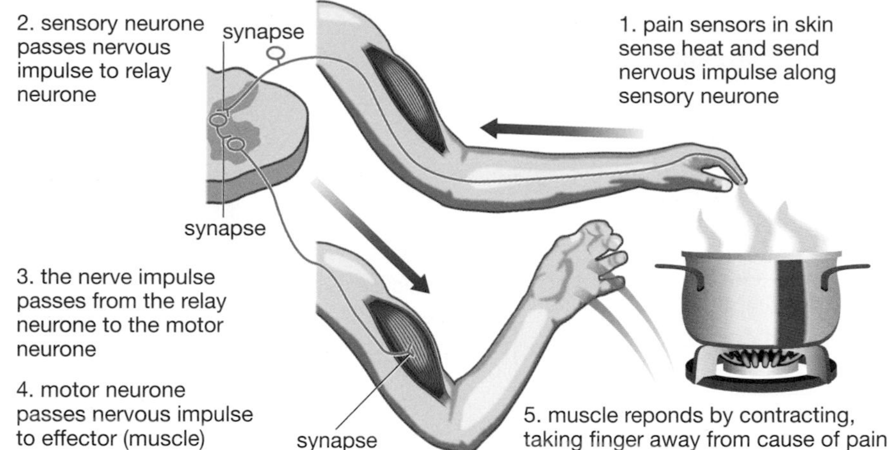

Fig. 2j.05: A reflex arc caused by touching a hot object

The simple arc allows a rapid automatic response, preventing damage.

Voluntary and involuntary responses

A **voluntary** response is one you consciously choose to do. It involves thought and so is under concious control by the brain cortex. It is a slower response than a reflex arc, but allows you to change how you respond (e.g. choosing to hold on to something very hot until you can put it down safely).

An **involuntary** response is one that happens automatically, as in a reflex action. It is usually a more rapid response than a voluntary one, and does not change. These responses help protect us from dangerous situations.

Antagonistic muscles

Muscles can only contract actively – to return to their original length, they must be pulled by something else. So muscles may be arranged in **antagonistic pairs**.

The antagonistic pair of biceps and triceps muscles in the arm causes the lower arm to lift and fall around the elbow joint.

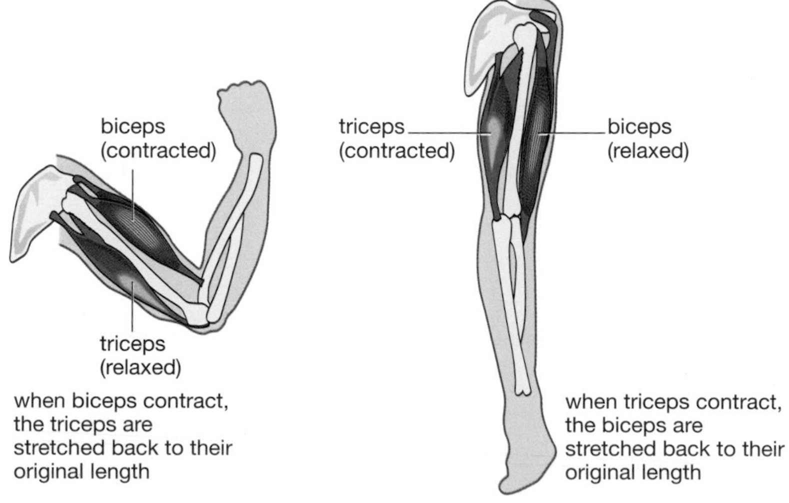

Fig. 2j.06: The biceps and triceps of the upper arm are antagonistic muscles

Structure and function of the eye

Many structures in the eye support the light-sensitive cells in sensing light.

Components	Function
retina	layer at back of eye containing light-sensitive cells
cornea	transparent surface where most light refraction occurs as it enters the eye
iris	made of radial and circular muscles that control the size of the pupil
pupil	hole in iris where light passes to the back of the eye
lens	transparent body that changes shape due to ciliary muscles, manages fine focusing of image on retina
optic nerve	carries nerve impulses from light-sensitive cells to brain

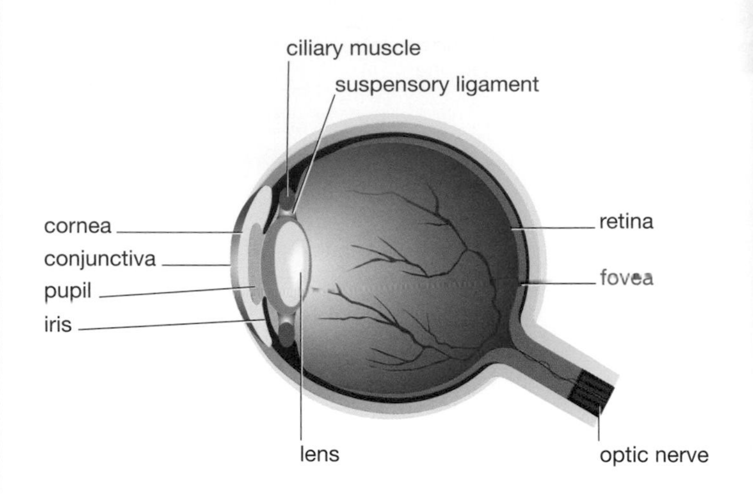

Fig. 2j.07: Components of the eye

CAM

Focusing on near and distant objects

The ciliary muscles control the thickness of the lens, so adjusting the fine focus of the image on the retina. This is known as **accommodation**.

When focusing on a near object:

- the light rays reaching the eye are diverging
- the ciliary muscles contract
- this reduces the pull on the suspensory ligaments attached to the lens
- the lens becomes fatter.

A fatter lens increases refraction of the light rays, to produce a clear image on the retina.

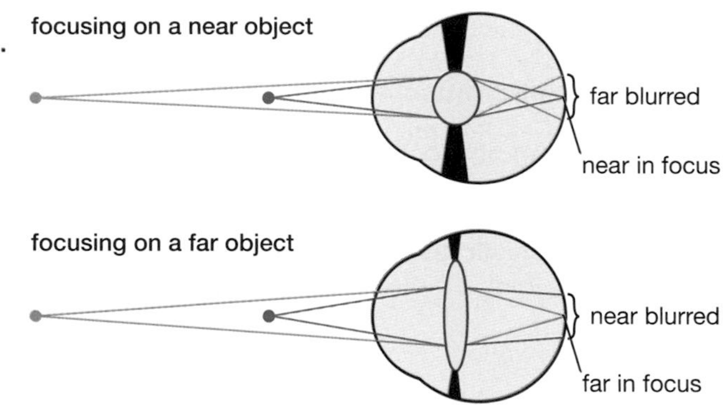

Fig. 2j.08: Focusing of the eye

When focusing on a distant object:

- the light rays are parallel
- the ciliary muscles relax
- this increases the pull on the suspensory ligaments attached to the lens
- the lens is pulled thinner
- the light rays are refracted less as they pass through the lens, and so focus a clear image on the retina.

Responding to changes in light intensity

The size of the pupil changes, depending on light intensity. In bright light it constricts (gets smaller) to protect the light-sensitive cells of the retina from damage by high light intensity. In dim light it dilates (gets wider) to allow more light in for better vision.

AM This is known as the **pupil reflex** because it is an automatic response.

The size of the pupil is controlled by circular and radial muscles.

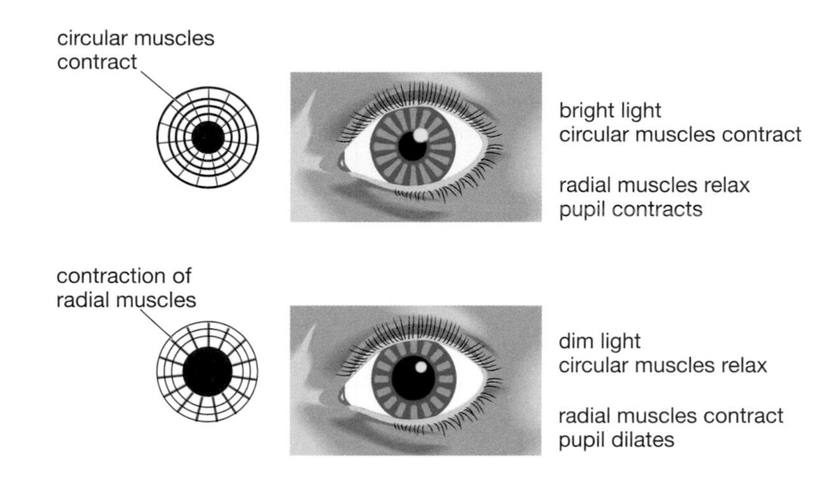

Fig. 2j.09: The pupil reflex to light intensity

CAM

Rods and cones

There are two types of light-sensitive cells in the retina.

Cell type	Distribution	Function
rod cells	more in the peripheral (outer) regions of the retina	• respond to light intensity • used mainly for vision in dim light
cone cells	most in centre of retina, especially fovea	• respond to one of three colours (red, green, blue) • for colour vision in bright light

Human hormones

Hormone	Source	Target organs	Effect
ADH (antidiuretic hormone)	pituitary	collecting ducts of kidney tubules	increases permeability so increases water reabsorbed in kidneys, more concentrated urine produced
adrenaline	adrenal glands (above kidneys)	many target organs, e.g. heart pupils liver and muscle cells	prepares body for action: increases heart rate dilate converts glycogen to glucose
insulin	pancreas	liver and muscle cells	removes glucose from blood and converts to glycogen, so reducing blood glucose concentration
testosterone	testes	sperm-forming cells many target organs at puberty	produces sperm causes secondary sexual characteristics (see Section 3A)
progesterone	ovaries	uterus	causes increased development of uterus lining control of menstrual cycle (see Section 3A)
oestrogen	ovaries	uterus many target organs during puberty	control of menstrual cycle and development of secondary sexual characteristics (see Section 3A)

Negative feedback

Negative feedback occurs in systems that are self-controlling. If a change happens in the system, a response by the system causes the opposite change to happen.

An important example in the body is the control of blood glucose concentration. This is monitored by cells in the pancreas, which detect if the concentration rises too high or falls too low.

- Digestion of food causes an increase in blood glucose concentration as glucose is absorbed from the small intestine.
- Cells in the pancreas detect the rise in blood glucose and cause the pancreas to secrete the hormone **insulin** into the blood.
- Insulin causes cells in the liver and muscles to remove glucose from the blood and convert it to glycogen for storage.
- If blood glucose falls too low, cells in the pancreas detect the fall and secrete a different hormone, **glucagon**, into the blood.
- Glucagon stimulates liver cells to break down glycogen to glucose and release it into the blood.

This negative feedback response keeps blood glucose concentration normally within very tight limits.

Hormones in food production

Hormones may be added to the food of animals used for food production, to stimulate the production of more of the food we eat. Oestrogen makes animals retain water and growth hormone makes animals grow quicker. For example:

- increased meat production in beef cattle
- increased egg laying in chickens.

Many people are concerned with the effects on us of eating food containing extra hormones, or on the environment if the hormones enter water supplies. So hormone use in food is tightly restricted, and banned in some countries.

Drugs

A **drug** is a substance taken into the body where it changes chemical reactions.

Medicinal drugs help to treat symptoms or illnesses, and are often prescribed by a doctor. For example, **antibiotics** that are used to kill or stop the growth of bacteria that are causing infection in our bodies.

Note that antibiotics work by affecting cell processes in bacteria. Antibiotics therefore cannot be used to treat viruses because they do not have the same cellular structures.

Other drugs are taken for their effect on the body, e.g. heroin, alcohol, tobacco smoke.

Heroin

This drug is taken for its powerful **depressant** effect, which reduces feelings of anxiety or pain. It causes many problems because:

- it is highly **addictive**, so the body quickly gets used to it and needs increasing amounts to produce the same effect
- it causes severe **withdrawal symptoms**, such as nausea, sweating, sleeplessness, shivering, if the body doesn't get enough
- as the need increases, so does the cost, so addicts often turn to crime to get the money they need to buy heroin
- the drug is injected, and addicts may share injection needles, which increases the risk of infection with blood-borne diseases such as HIV (human immunodeficiency virus) that causes AIDS (acquired immunodeficiency disease).

Alcohol

Alcohol is also a depressant, so many people take it for its relaxing effect. Small amounts of alcohol increase reaction times which is dangerous when driving.

In excessive amounts, alcohol causes:

- reduced self-control, which may lead to extreme behaviour including violence to others
- vomiting, as the body tries to get rid of the toxic effects of alcohol
- unconsciousness, increasing the risk of death through suffocation from vomit
- over a long period of excessive intake, damage to the liver.

Fig. 2j.10: Driving under the influence of alcohol is one of the most common causes of road deaths

Tobacco smoke

Smoking tobacco can have a calming effect. However, tobacco smoke contains dangerous chemicals:

- tar is a thick sticky substance that coats lung surfaces, reducing the area for gas exchange; it also blocks the action of cilia, increasing the risk of infections in the lungs
- nicotine is an addictive stimulant, making smoking difficult to give up
- carbon monoxide is a toxic gas that replaces oxygen in haemoglobin, so there is less oxygen in the blood – in excessive amounts this may cause cell death; in pregnant women the gas passes to the fetus, reducing its rate of growth, which can cause problems at birth and in later development
- smoke particles cause a lot of coughing which damages the delicate alveoli and reduces the area for gas exchange.

Other chemicals in smoke may cause cancers, such as lung cancer.

You should now be able to:

★ define the terms *stimulus*, *response* and *effector* (see pages 92–93)

CAM ★ identify the human sense organs that respond to: light, sound, touch, temperature, chemicals (see page 93)

★ give two examples of homeostasis in the human body (see pages 93–94)

★ describe the geotropic responses of roots and stems (see page 94)

★ describe positive phototropism of stems (see page 95)

★ compare the nervous and hormonal communication systems in humans (see page 96)

CAM ★ describe the functions and adaptations of three types of neurone (see page 96)

★ describe a reflex arc (see page 97)

CAM ★ compare voluntary and involuntary actions (see page 97)

★ explain why muscles may be arranged in antagonistic pairs (see page 97)

★ describe how the structure of the eye is adapted to its function of responding to light (see page 98)

CAM ★ describe the response of the eye in focusing near and distant objects (see page 98),

★ describe the response of the eye to changes in light intensity (see page 99)

★ describe how changes in the skin help to regulate body temperature (see pages 93–94)

CAM ★ identify structures in skin that are associated with body temperature regulation

★ identify the sources, roles and effects of: ADH, adrenaline, insulin, testosterone, progesterone and oestrogen (see page 100)

CAM ★ describe the negative feedback control of blood glucose concentration (see page 100)

★ describe a problem with using hormones in food production (see page 101)

★ explain what we mean by the term *drug* (see page 101)

★ explain why antibiotics can be used against infections caused by bacteria but not by viruses (see page 101)

★ describe problems caused by the use of heroin (see page 101)

★ describe problems caused by alcohol (see page 101)

★ describe problems caused by smoking tobacco (see page 102).

Practice questions

1.

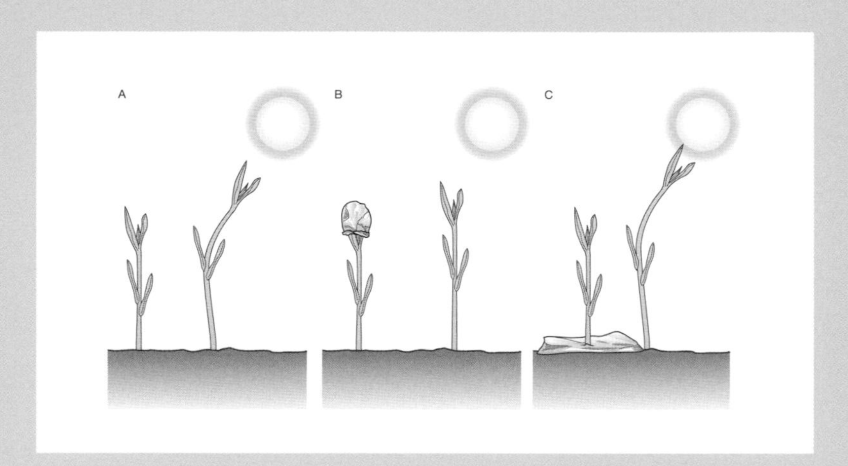

(a) What region of the plant shoot is sensitive to light? **(1)**

(b) Describe the evidence of the practical that supports your answer to the above. **(2)**

(c) Explain what causes the plant stem to bend towards the light. **(3)**

(d) Why is this response of the shoot to the direction of light beneficial to the plant? **(2)**

2.

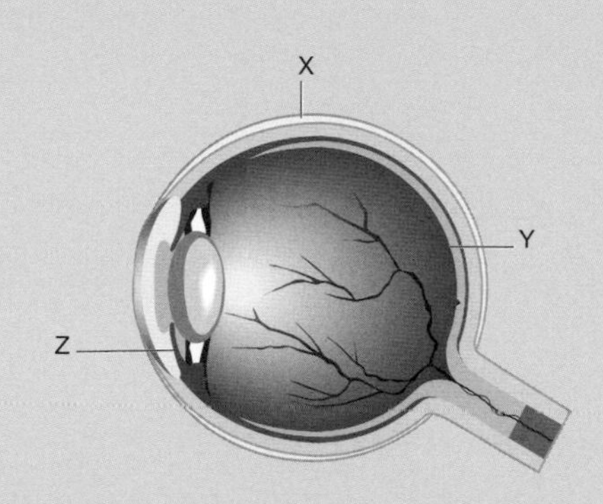

(a) Name the parts of the eye labelled X, Y and Z. **(3)**

AM (b) On a copy of the diagram above label the part of the eye that has the most cone cells C. **(1)**

(c) Name the two parts of the eye that focus light on the retina A and B and label them on your diagram. **(2)**

(d) On your diagram label the optic nerve O. If O were cut, why would a person be blind in that eye? **(2)**

AM (e) When you look out of the window, then look back at this question to read it, the shape of the lens in your eyes changes. Explain how. **(4)**

3. (a) State how hormonal control differs from a reflex action, referring to:

(i) how the message is transmitted **(1)**

(ii) the speed of transmission **(1)**

(iii) how long the effect lasts. **(1)**

(b) Name the parts of the nervous system involved in a reflex arc, starting with the stimulus. **(4)**

(c) What is a synapse? **(3)**

Section Three

3 Reproduction and inheritance

A Reproduction

You will be expected to:

- ★ describe the differences between sexual and asexual reproduction
- ★ define the term *fertilisation*

Flowering plants

- ★ describe how wind-pollinated and insect-pollinated flowers are adapted for pollination
- ★ describe how pollination leads to fertilisation in a plant
- CAM ★ compare self-pollination and cross-pollination
- ★ describe the conditions needed for seed germination
- ★ describe how a germinating seedling uses food reserves
- ★ describe how plants can be reproduced asexually
- CAM ★ describe examples of seed dispersal by wind and by animals

Humans

- ★ name the structures of the male and female reproductive systems in humans and describe their functions
- ★ describe the role of oestrogen and progesterone in the menstrual cycle
- CAM ★ describe the role of LH and FSH in the menstrual cycle
- ★ define the terms *growth* and *development*
- ★ outline the development of the fetus, and the processes of labour and birth
- ★ describe the antenatal care of women
- ★ describe the role of the placenta and amniotic fluid
- ★ state the role of oestrogen and testosterone in the development of secondary sexual characteristics
- CAM ★ compare methods of birth control
- ★ describe the symptoms and treatment of gonorrhoea
- ★ outline the effects of HIV and describe how its transmission can be prevented.

Sexual and asexual reproduction

Reproduction is the production of new individuals of the same species.

The differences between **sexual reproduction** and **asexual reproduction** are shown in the table.

	Asexual reproduction	Sexual reproduction
time taken to produce offspring	relatively fast	relatively slow
number of parents	one	two
genetic variation	none – offspring are clones of parent and of each other (genetically identical)	each offspring is genetically unique (different from each other and from parents)
gametes	none	male and female
cell processes involved	• mitosis (division of body cell)	• meiosis to form gametes • fertilisation of female gamete by male gamete • mitosis
where offspring grow	directly from body of parent, dispersal limited	for external fertilisation, can be far from parents

CAM

Asexual reproduction is advantageous when:

- there is no individual of the opposite sex to mate with
- environmental conditions are not changing
- reproduction needs to be quick (such as during the summer).

Sexual reproduction is advantageous when:

- environmental conditions are changing (providing variation in offspring).

Examples of asexual reproduction

- Bacteria – reproduce asexually by simple cell division called binary fission to form two identical daughter cells. Under optimum conditions, binary fission can occur every 12–20 minutes. (*Not called mitosis as there is no nucleus and only one chromosome.)
- Fungi – spores formed by asexual reproduction are the dispersal stage of fungi, produced in fruiting body structures, such as the pinheads on moulds. Fungi also reproduce asexually by fragmentation of the mycelium.
- Plants – such as the production of tubers in the potato *Solanum tuberosum.* Food is stored in the tuber over winter, while the rest of the plant dies. In the spring the food stores are used to grow new plants. (See also runners, on page 111.)

Fig. 3a.01: Fungal spores are dispersed by wind when it is dry

Sexual reproduction in plants

Fertilisation involves the fusion (joining together) of the **haploid** nucleus from a male gamete with the haploid nucleus of a female gamete.

This produces a **zygote,** which is a **diploid** cell. The zygote divides by cell division (mitosis, see Section 3B) to form an **embryo**.

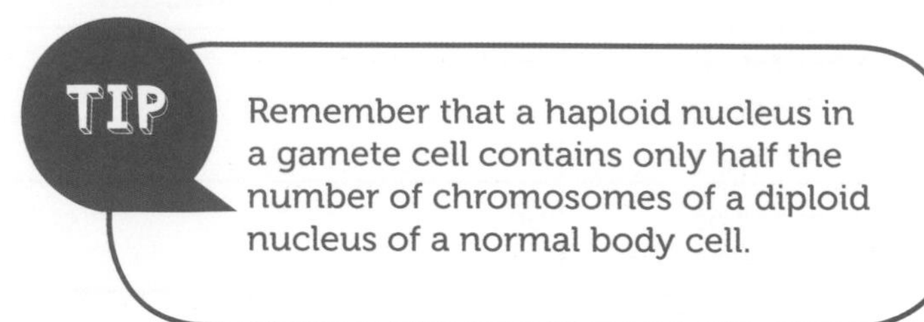

The structure of flowers

The flower is the reproductive structure of flowering plants. Some flowers are male-only or female-only, but others contain both male and female structures.

- Male structures include the **stamens**, **anthers** and **pollen grains** containing the male nuclei.
- Female structures are within the **carpel**, consisting of the **stigma**, **style** and **ovaries**, which contain the female nuclei.

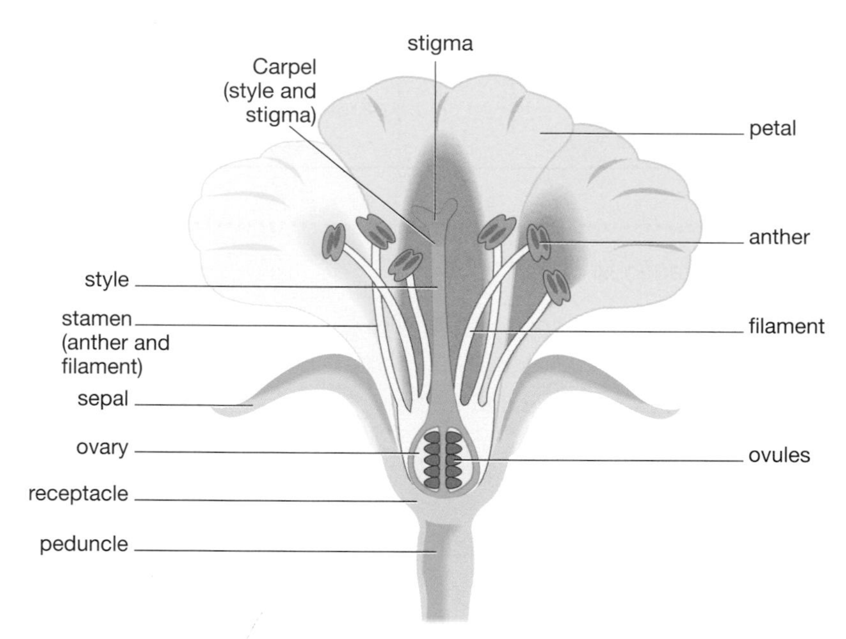

Fig. 3a.02: Structure of a flower

CAM

The **sepals** protect the flower bud before it opens and may be used to attract pollinators. The **petals** may be developed to attract pollinators.

Pollination

Pollination is the transfer of pollen from the anther of one flower to the stigma of another flower. It must occur before fertilisation can take place.

Many flowering plant species depend on insects to transfer the pollen from one flower to another. **Insect-pollinated flowers**:

- attract insects with colours and/or scent and/or food rewards such as nectar and protein-rich pollen
- have pollen grains that are relatively large and may be covered in spines to stick to the insect's body.

Other flowering plants have **wind-pollinated flowers** that:

- produce large quantities of pollen
- have lightweight pollen grains that are tiny and smooth, and may have structures that increase wind-resistance and therefore the time they can stay in the air
- have anthers and stigmas that dangle outside the flower to distribute and catch the pollen.

Fig. 3a.03: (Left) Insect-pollinated flower and (Right) wind-pollinated ragweed

CAM

If pollen is transferred from a flower on one plant to a flower on a different plant of the same type it is called **cross-pollination**.

If pollination occurs in the same flower, or on a flower of the same plant, it is called **self-pollination**.

Self-pollination is an example of sexual reproduction that only involves one parent, but because gametes are formed by meiosis and fertilisation occurs, there is still genetic variation in the seeds.

More genetic variation occurs with cross-pollination as gametes from two different parent plants fuse.

Fertilisation

After pollination, the pollen grain grows a **pollen tube** from the stigma down through the style to the ovary.

The male gamete nucleus in the pollen grain moves down the pollen tube to fuse with the nucleus of the female gamete in the ovary.

After fertilisation:

- a plant **embryo** forms by mitosis and cell differentiation
- stored nutrients form **cotyledons** around the embryo
- the embryo and food stores are surrounded by a tough or hard layer (**testa** or seed coat) to form a **seed**
- the ovary wall may develop to form a **fruit** as extra protection around the seed.

The fruit may be fleshy, e.g. in berries, or dry, e.g. the shell of a nut.

CAM

Structure of a seed

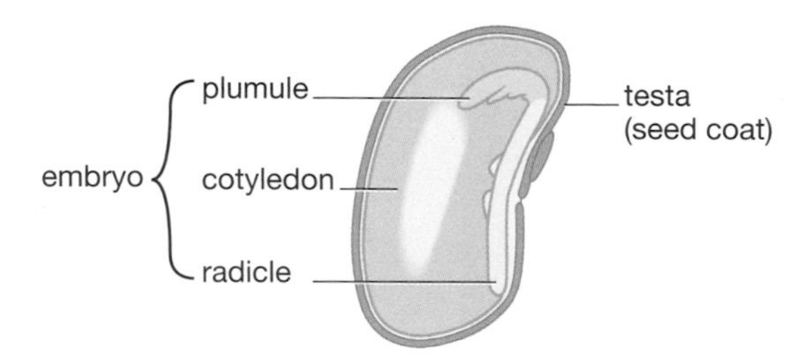

Fig. 3a.04: Section through a bean seed; the cotyledons hold the food stores; it has very little endospermic tissue

The **plumule** of the embryo will grow and develop to form the shoots and leaves of the new plant.

The **radicle** of the embryo will grow and develop to form the roots of the new plant.

The food stores can be tested with iodine solution to show the presence of stored starch.

Seed dispersal

Seeds are dispersed away from the parent plant to avoid overcrowding and competition between parent and seedlings.

The seeds may be dispersed by:

- animals when eaten, e.g. fleshy fruits eaten, seeds pass through alimentary canal undigested and are deposited in animal's faeces
- animals, by attaching to the body, e.g. burdock burs, and dropping off later
- wind – structures help increase air resistance time that seed is carried by wind, e.g. feathery plumes on a dandelion seed, and 'wings' on sycamore seed
- water – seeds have tough coat to prevent rotting by water, and are light enough to float, e.g. coconut
- explosive dispersal – seeds are flung from the fruit as it breaks open.

Fig. 3a.05: Seed dispersal from a dandelion seed

Germination

Seeds may remain dormant until conditions for germination are right. To germinate, seeds need:

- water absorbed into cells to mobilise enzymes so metabolic reactions can occur
- oxygen for respiration
- to be at an optimum temperature for enzyme activity.

The food stores (cotyledons) in the seed provide food for respiration. This releases the energy needed for growth and differentiation until chlorophyll-containing tissue (e.g. young leaves) develops and can start to photosynthesise.

Asexual reproduction in plants

Natural asexual reproduction

Plants such as the strawberry can reproduce asexually by horizontal stems called **runners**. These grow along the ground and develop roots at stem joints. Buds on the runner develop into new strawberry plants. When the root system is well developed, the runner dies, separating the daughter plant from the parent plant.

Artificial asexual reproduction

Plant growers may use artificial methods of asexual reproduction to grow more plants from one parent plant, e.g. by taking **cuttings** of stems, roots or leaves.

- A piece of plant, e.g. a shoot, is cut from a parent plant.
- The end of the cutting may be dipped in plant hormones to encourage root growth and development.
- The cutting is usually placed in moist soil or compost.
- After a few days new roots start to grow, and eventually new shoots and leaves grow.

The human male and female reproductive systems

The key structures in the human male reproductive system are the two **testes**, the sperm ducts, the prostate gland (and seminal vesicle) and penis.

Sperm passes out of the body through the urethra in the penis which, at other times, carries urine from the bladder out of the body.

The testes are held outside the body in a sac called the **scrotum**, so that they are at a lower temperature than the body because heat reduces sperm production.

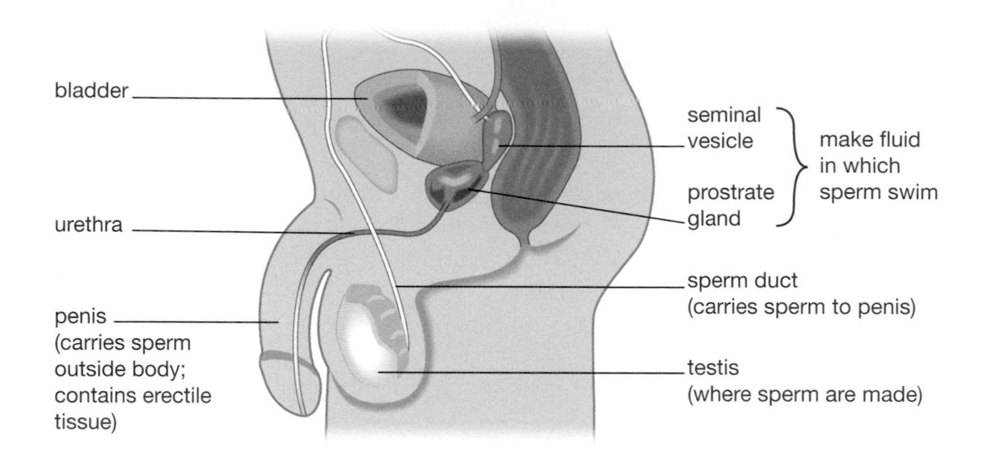

Fig. 3a.06: The human male reproductive system

Erectile tissue in the penis produces an erection, which enables the penis to be placed in the woman's vagina during sexual intercourse.

The key structures of the human female reproductive system are the ovaries, oviducts, uterus, cervix and vagina.

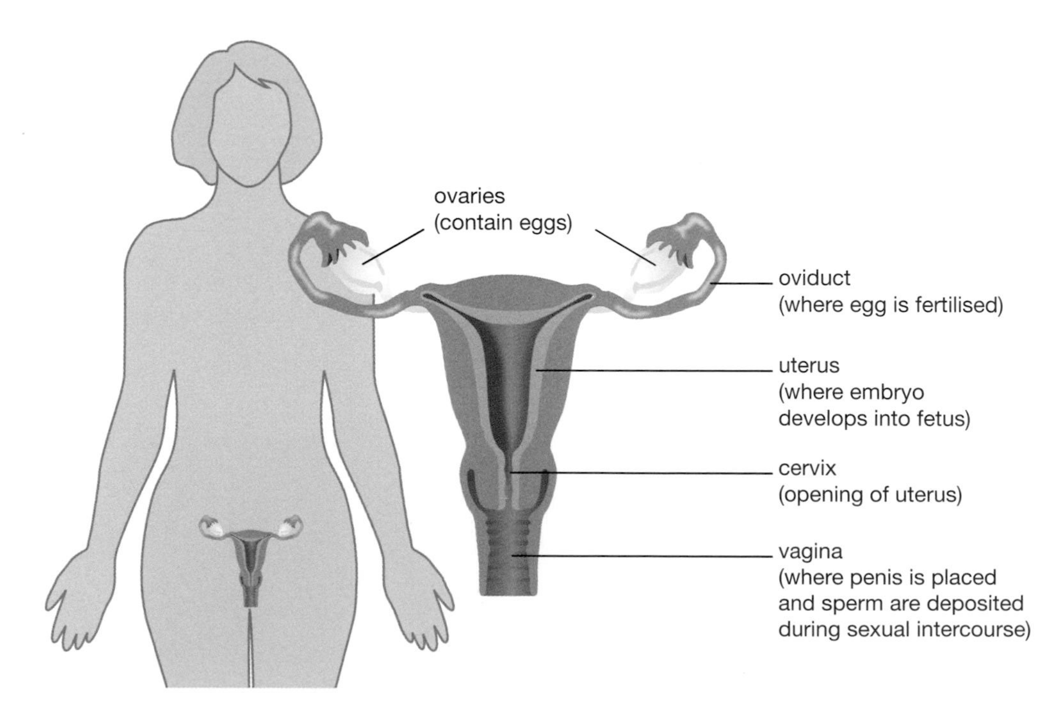

Fig. 3a.07: The human female reproductive system

Male and female gametes

More than 4 million sperm cells are released each time a man ejaculates.

Sperm cells are tiny compared to an egg, with a head containing the nucleus, and a tail so that they can swim.

A woman is born with about 200 000 eggs in each ovary. She does not produce any more eggs after birth, but usually releases one mature egg each month when she ovulates.

The human egg cell is one of the largest of the human cells, at about 0.2 mm in diameter, and contains food stores. It cannot move on its own. It moves down the oviduct by the action of ciliated cells in the oviduct.

The menstrual cycle

The **menstrual cycle** is the cycle of development of the uterus lining, **ovulation** (release of an egg from the ovary) and breakdown of the uterus lining followed by a period, which happens about every 28 days in women from puberty to the age of about 50.

The menstrual cycle is controlled by hormones.

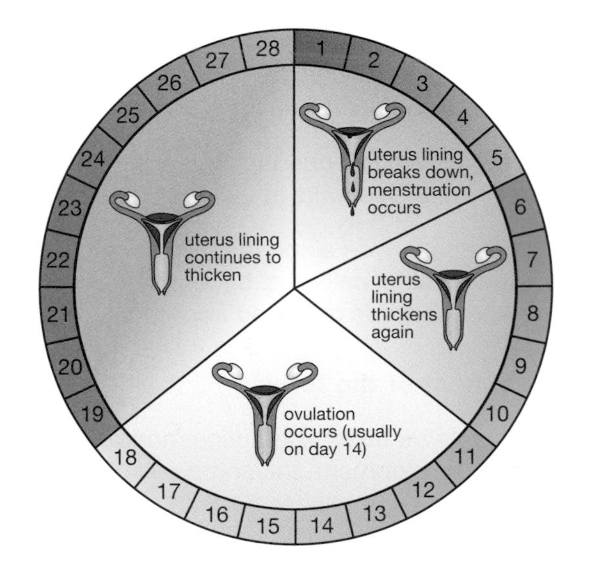

Fig. 3a.8: The menstrual cycle

Oestrogen

Oestrogen is produced by the ovaries.

- It is produced in increasing quantities in the first 14 days of the cycle, resulting in ovulation.
- During the second half of the cycle, it works with progesterone to cause increased thickening of the uterus lining.

If the egg is not fertilised, production of oestrogen decreases towards the end of the cycle. If the egg is fertilised, oestrogen levels remain high to maintain the uterus lining.

Progesterone

Progesterone is also produced by the ovaries.

- It is produced in increasing quantities in the second half of the cycle.
- With oestrogen, it causes increased thickening of the uterus lining.

If the egg is not fertilised, production of progesterone decreases at the end of the cycle, leading to breakdown of the uterus lining and the period.

If the egg is fertilised, progesterone levels remain high to maintain the uterus lining.

After the egg is released from the follicle, the empty follicle is called the *corpus luteum*. The corpus luteum secretes oestrogen and progesterone if fertilisation occurs.

LH and FSH

The pituitary gland at the base of the brain secretes two other hormones that help to control the menstrual cycle.

- **LH** (luteinising hormone) is produced in large quantities during days 12–14 as a result of high levels of oestrogen, and triggers ovulation.
- **FSH** (follicle-stimulating hormone) is produced mainly in the first few days of the cycle and causes egg cells in the ovaries to mature, ready for one to be released during ovulation.

Oestrogen and progesterone inhibit the production of FSH, so maturing of another egg can only occur when oestrogen and progesterone levels fall at the end of the cycle if fertilisation has not occurred.

Sexual intercourse and fertilisation

At ovulation, the released egg moves along the oviduct by the action of cilia on the inside of the tube, towards the uterus.

During sexual intercourse the man inserts his erect penis into the vagina of the female, and sperm are released near the cervix during ejaculation.

Sperm cells swim through the cervix, into the uterus and up the oviducts. Fertilisation (fusion of the male and female nuclei) occurs in the oviduct to form a zygote.

The zygote continues along the oviduct into the uterus. During this time the cell divides by mitosis, producing a ball of cells that becomes the **embryo**. The embryo implants into the thickened uterus lining, where it gets oxygen and food from the mother until the placenta develops.

Development of the fetus

The human fetus takes about nine months to develop in the uterus before it is born. After about three months of development, the embryo is called a **fetus**.

- During the fourth week the heart begins to beat.
- After eight weeks, the limbs have begun to form and the placenta starts to develop.
- At the end of the first three months, all organs are present. Up until this point, the developing embryo is vulnerable to drugs, chemicals and radiation that can cause birth defects.
- From about the fourth month, fingers, toes and facial features are well formed, limbs get longer and can move.
- During the last three months the organs mature and function – liver stores glycogen, kidneys produce urine, brain shows activity cycles of sleep and waking, lungs show breathing movements.
- During the seventh month the fetus usually moves so that the head is pointing down, ready for birth.

Amniotic fluid

The fetus floats in and is protected by **amniotic fluid** in the amniotic sac. This liquid protects the fetus from physical bumping as the mother moves, and minimises changes in temperature.

The placenta

After about eight weeks, the **placenta** develops. This tissue is closely attached to the uterus wall and is connected to the fetus by the **umbilical cord**.

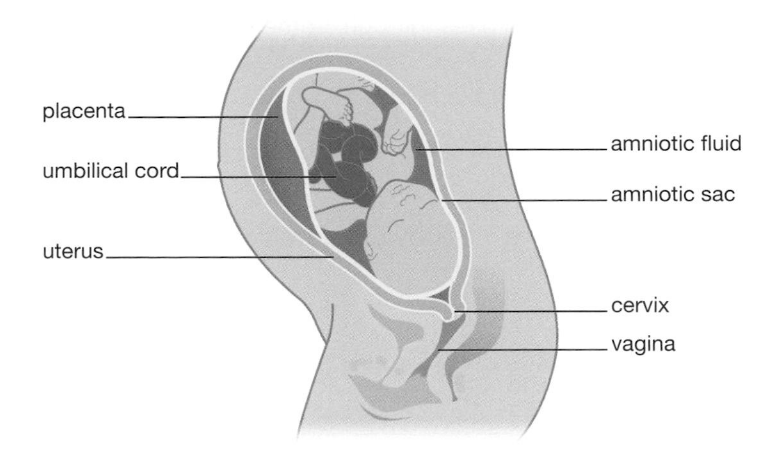

Fig. 3a.09: The fetus ready for birth

The fetus's blood goes through blood vessels in the umbilical cord to the placenta.

In the placenta, substances are exchanged:

- oxygen, glucose, antibodies and water move into the fetus's blood from the mother's blood
- carbon dioxide and urea move from the fetus's blood into the mother's blood.

The placenta is the source of all nutrients for the fetus until birth. The mother's blood does not mix with the fetus' blood.

Antenatal care

During pregnancy, the woman receives **antenatal care**. This includes:

- checks on the mother's health and the health and development of the fetus
- advice on how to improve health, such as giving up smoking and reducing alcohol intake
- advice on diet to supply sufficient nutrients, including minerals and vitamins, without eating too much
- advice on dietary supplements, including folic acid and possibly vitamin D, to avoid health risks in the fetus
- advice on exercises for fitness and preparation for birth
- advice on what to expect during labour and birth
- advice on caring for the baby after birth, including breast-feeding.

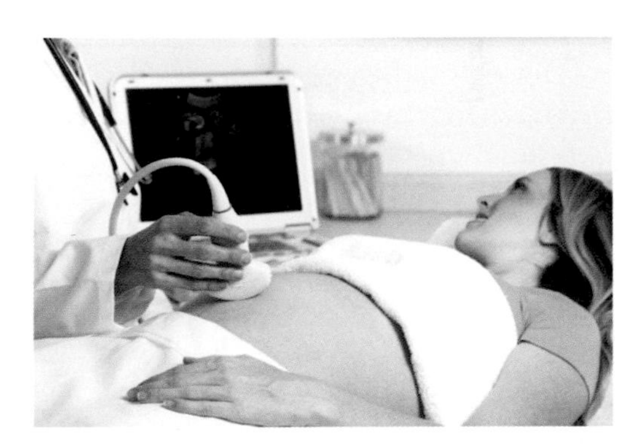

Fig. 3a.10: An ultrasound scan checks on the development of the fetus

Labour and birth

Labour is the sign that birth is about to happen.

- Contractions of the uterus wall become strong, to help push the fetus out.
- The cervix dilates, to allow the fetus to pass through, sometimes releasing a plug of mucus which is a sign of labour.
- The amniotic sac bursts, releasing amniotic fluid.

After the baby is born, the placenta, umbilical cord and amniotic membranes (together called the 'afterbirth') are pushed out.

The umbilical cord is clamped and cut and the baby can develop independently.

Feeding the baby

Babies may be breast-fed, or fed by bottle using formula milk made up with water.

	Breast-feeding	Bottle-feeding
Advantages	• usually perfect balance of nutrients for healthy development of baby • nutrient balance naturally changes as baby grows to support changes in growth and development • contains antibodies that protect against infection	• can be done by anyone, so can help father bond with baby, or give mother more freedom (e.g. if working), or if mother is ill or unable to feed baby
Disadvantages	• can only be done by mother • can be source of infection for baby, e.g. HIV virus, if mother is infected	• not a perfect balance of nutrients which can cause problems for some babies • does not contain antibodies

Secondary sexual characteristics

Secondary sexual characteristics develop in boys and girls at **puberty**. These changes are produced by increased secretion of sex hormones: oestrogen in girls, testosterone in boys. They prepare the body for sexual reproduction and make it obvious to others that the body is physically mature enough for sexual reproduction.

Secondary sexual characteristics in men:

- increased muscle mass
- deepening of voice
- development of hair on face, armpits and pubic region
- increased secretion of body odour.

Secondary sexual characteristics in women:

- development of breasts (for breast-feeding)
- widening of hips (for pregnancy and childbirth)
- development of hair on armpits and pubic region
- menstrual cycle
- increased secretion of body odour.

CAM

Birth control

Birth control allows a couple to have sexual intercourse without resulting in fertilisation. This helps a couple decide when to have children and control how many children they have.

Natural methods of birth control include:

- **abstinence** – not having sexual intercourse (e.g. before marriage)
- **rhythm method** – avoiding sexual intercourse during the time of ovulation, when fertilisation is most likely.

The rhythm method is not very reliable because sperm can remain active in the woman's body for up to two days after intercourse.

Chemical methods of birth control include:

- the **contraceptive pill** which contains female sex hormones, usually progesterone with some oestrogen, that interrupt the menstrual cycle and prevent ovulation – must be taken every day, or can be given as an implant under the skin that lasts for a few months
- **spermicide** – a chemical that kills sperm, most effective if used with a mechanical barrier.

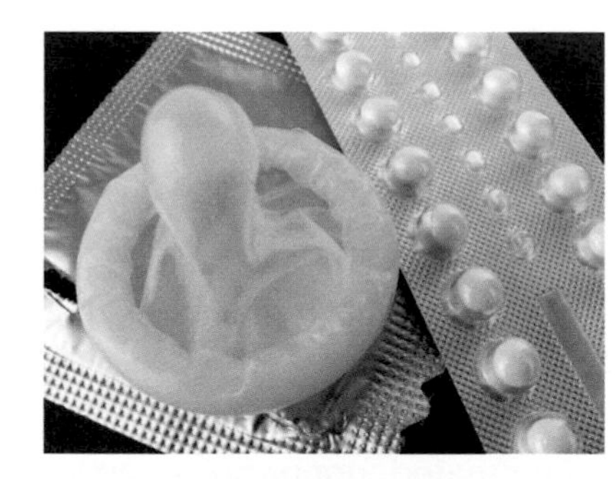

Fig. 3a.11: Two birth control methods

Most mechanical methods of birth control prevent the sperm from entering the woman's uterus:

- **condom** – a thin rubber sheath placed over the erect penis before ejaculation (also good protection against sexually transmitted diseases)
- **diaphragm** – a rubber or plastic cap placed over the entrance of the uterus before sexual intercourse
- **femidom** – a female equivalent of the condom that is placed in the vagina before sexual intercourse.

The **IUD** (intra-uterine device) is a mechanical device placed in the uterus of a woman by a doctor or nurse. It can remain in place for several years, and prevents pregnancy by preventing sperm reaching the egg, or by preventing the zygote from implanting in the uterus wall.

Surgical methods of birth control prevent the gametes from moving to where fertilisation can occur:

- A **vasectomy** cuts the sperm ducts so the sperm cannot get to the urethra and leave the body.
- In **female sterilisation**, the oviducts are blocked or cut, so the eggs cannot get to the uterus.

Surgical methods are difficult to reverse and so are most useful for couples who do not want more children.

Artificial insemination and fertility drugs

Some couples who would like to have children are unable to do so.

If a man has very few sperm, or the sperm do not swim well, fertilisation is unlikely to happen naturally. A doctor can take the sperm from the man, select the best or make them more active, then place the sperm in the woman's uterus by **artificial insemination**, to increase the chance of fertilisation.

If a man has no sperm, artificial insemination using a sperm donor (from another man) can be used to fertilise the woman's egg.

If a woman has difficulty releasing eggs from her ovaries, **fertility drugs** that include the sex hormones can be used to encourage more eggs to be released.

There are social implications of using these methods.

- Couples who were unable to have children may be able to have their own family.
- The treatment may be too expensive for some couples (although some treatments are free in the UK via the NHS).
- A child born as the result of sperm donation may want to know who is their biological father.
- Fertility drugs may cause several eggs to be released at the same time, resulting in a multiple birth which can be a risk to the mother's health during pregnancy, and to the health of the babies after birth due to low birth weight.

Sexually transmitted diseases

Sexually transmitted diseases are diseases transmitted during sexual intercourse, usually by the exchange of body fluids such as semen and vaginal mucus.

Gonorrhoea

Gonorrhoea is caused by a bacterium. Symptoms of infection may include:

- burning sensation when urinating
- foul-smelling mucus discharged from the urethra or vagina.

However, many infected people show no symptoms.

If untreated, gonorrhoea can lead to sterility, and affect the heart and joints. Treatment is with a course of antibiotics.

HIV

HIV (human immunodeficiency virus) is transmitted in many body fluids:

- during sexual intercourse with an infected sexual partner
- during pregnancy, from infected mother to fetus across the placenta
- during breast-feeding, from infected mother to baby in breast milk
- sharing with an infected person the needles used to inject drugs such as heroin
- by a blood transfusion or organ transplant from an infected person.

HIV in the blood attacks the immune system, so other pathogens can infect the body and cause illness. If untreated, HIV can result in the disease AIDS (acquired immunodeficiency syndrome) which can lead to severe illness and death.

The spread of HIV can be limited by:

- restricting sexual intercourse to partners who are not infected
- preventing the exchange of body fluids during sexual intercourse by using a condom or femidom
- using sterile needles for drug injection
- screening of all blood for transfusion and organs for transplant
- bottle-feeding of a baby by an infected mother.

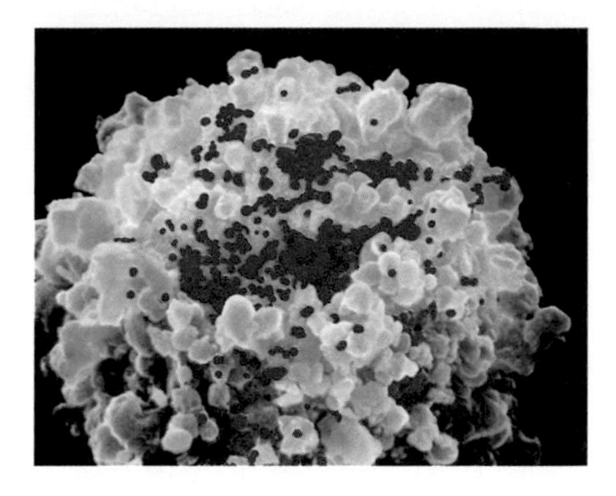

Fig. 3a.12: The HIV virus

Growth and development

Growth is a permanent increase in size and dry mass by an increase in cell number or cell size or both.

Development is an increase in complexity of an organism, such as when cells differentiate into different types to form different tissues and organs.

You should now be able to:

★ describe the differences between sexual and asexual reproduction (see page 107)
★ define the term *fertilisation* (see page 108)
★ describe the adaptations of an insect-pollinated and a wind-pollinated flower (see page 109)
★ describe how pollination leads to seed and fruit formation (see page 110)
CAM ★ describe self-pollination and cross-pollination (see page 109)
★ give examples of how seeds are dispersed (see page 110)
★ describe the conditions for seed germination (see page 111)
CAM ★ describe the structure of a seed in a fruit (see page 110)
★ describe how plants can reproduce asexually, naturally and by artificial methods (see page 111)
★ describe the structure and function of the human male and female reproductive systems (see page 112)
★ describe the roles of oestrogen and progesterone in the menstrual cycle (see page 113)
CAM ★ explain the roles of FSH, LH, oestrogen and progesterone in the menstrual cycle (see page 114)
★ compare male and female gametes in terms of size, numbers and mobility (see page 113)
★ outline the development of the zygote and fetus (see pages 114–116).

- ★ describe the role of the placenta (see page 115)
- ★ describe the function of amniotic fluid (see page 114)
- CAM ★ describe the antenatal care of pregnant women (see page 115)
- ★ describe the processes involved in labour and birth (see page 116)
- ★ compare the advantages and disadvantages of breast-feeding and bottle-feeding using formula milk (see page 116)
- ★ name the hormones that control the development of secondary sexual characteristics (see page 117)
-  ★ outline the following methods of birth control: natural, chemical, mechanical, surgical (see pages 117–118)
- ★ describe issues in the use of artificial insemination and fertility drugs (see page 118)
- ★ describe the symptoms, signs, effects and treatment of gonorrhoea (see page 118)
- ★ describe the ways in which HIV/AIDS can be prevented from spreading (see pages 118–119)
- ★ outline how HIV affects the immune system (see page 119)
- ★ define the terms *growth* and *development* (see page 119).

Practice questions

1.

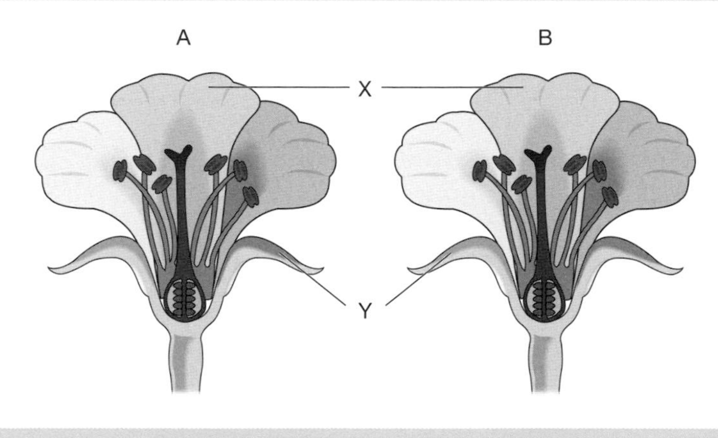

Plants A and B are the same species and they each produce flowers.

(a) State the functions of the parts labelled X and Y. **(2)**

(b) Mr Burns wants to cross plant A with plant B.

(i) Name the part of plant A from which he will get the pollen. **(1)**

(ii) Name the part of plant B on to which he will brush the pollen. **(1)**

(c) Describe what the pollen grain does after pollination. **(2)**

(d) What kind of cell division took place to produce the pollen? **(1)**

(e) What kind of cell division takes place after the ovule has been fertilised? **(1)**

2.

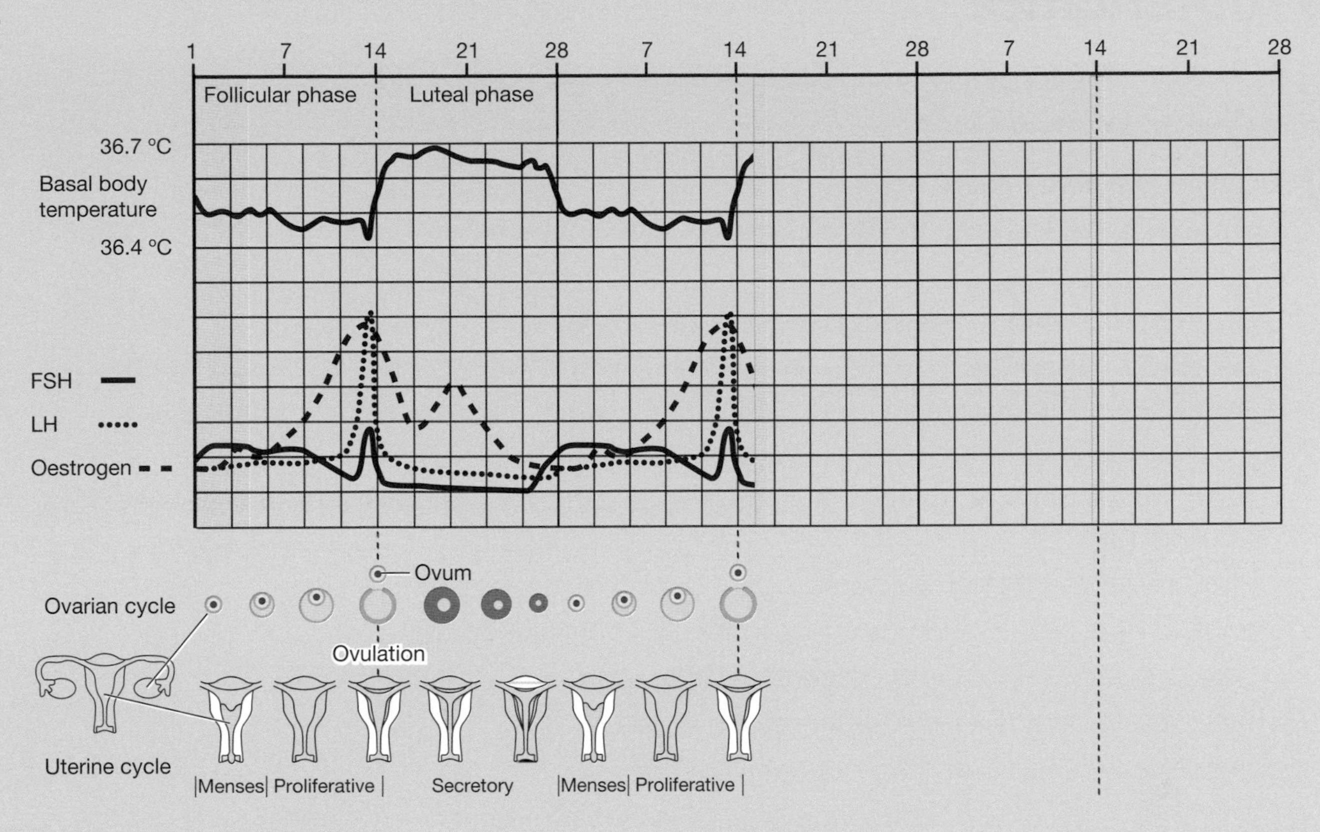

(a) State what happened between day 1 and day 4 in the first cycle. **(1)**

(b) Explain why the thickness of the uterus lining changes in the way shown in the graph from day 4 until day 28 of the first cycle. **(1)**

(c) On a copy of the diagram above, complete the graph from day 44 (day 16 of cycle 2) to day 60 (day 16 of cycle 3) to show what happens to the body temperature, the levels of FSH, LH and oestrogen, to the thickness of the uterine lining and the corpus luteum (follicle from which the egg was released) if fertilisation occurs. Explain why these changes are important after fertilisation has occurred. **(3)**

CAM

(d) What kind of cell division formed the sperm and egg? **(1)**

(e) What kind of cell division does the zygote undergo? **(1)**

(f) If a woman has blocked oviducts, she cannot have her own baby naturally. Suggest how her eggs could be removed for fertilisation. **(2)**

(g) Her eggs can be mixed with sperm in a petri dish, and after a few days the zygotes can be put back in the woman's uterus.

(i) Why are the sperm mixed with the eggs before being put back in the woman's uterus? **(2)**

(ii) Explain why the zygotes are kept for a few days before being put back in the woman's uterus. **(1)**

(h) Give two reasons why 'test tube baby' is not an accurate description of the above technique. **(2)**

(i) List three changes that take place in the uterus to help protect and nourish the developing embryo. **(3)**

Material in **blue** to be tested in CIE only

B Inheritance

You will be expected to:

- ★ define the term *inheritance*
- ★ state that the nucleus contains chromosomes that are formed from a string of genes
- ★ define the term *gene*
- ★ describe the structure and form of a DNA molecule
- ★ define the term *allele*
- ★ define the terms *dominant, recessive, homozygous, heterozygous, phenotype, genotype* and *codominance*
- ★ describe patterns of monohybrid inheritance using diagrams and predict probabilities of outcomes
- ★ interpret family pedigrees
- ★ explain how sex is inherited through the sex chromosomes
- ★ describe the function and process of mitosis
- ★ describe the function and process of meiosis
- ★ describe how fertilisation produces variation
- ★ give examples of variation that are genetic, environmental or a combination of both
- ★ define and give examples of *continuous variation* and *discontinuous variation*
- ★ define the term *mutation* and explain how it may be harmful, neutral or beneficial
- ★ describe evolution through natural selection
- ★ explain evolution of antibiotic resistance in bacteria
- ★ name factors that increase the incidence of mutations.

Definitions in inheritance

Inheritance can be defined as the transmission of genetic information from one generation to the next, through reproduction.

- The **nucleus** of a cell contains chromosomes.
- A **haploid nucleus** contains only one set of chromosomes – a copy of one of the sets of chromosomes in the diploid cell from which it was formed. Gametes (e.g. sperm and egg cells) are haploid. In humans, the number of chromosomes in each gamete is 23.
- A **diploid nucleus** contains two sets of chromosomes – one set from the male parent and one from the female parent. All body cells are diploid. In humans, the number of chromosomes in a diploid cell is 46.
- Each **chromosome** is a molecule of **DNA** (deoxyribonucleic acid) that carries a string of units of genetic information called **genes**. Genes are therefore sections of a molecule of DNA.

Each gene codes for a specific protein molecule. The protein can directly determine a characteristic or work with other proteins coded for other genes to determine a characteristic. The genetic information of an organism is stored in the DNA. The DNA, and therefore the genes, are copied and passed to the offspring (next generation) during reproduction.

- A gene can exist in more than one form called **alleles**, e.g. the gene for colour vision can either be the normal colour vision allele or the colour blind allele. Since diploid cells have two sets of chromosomes, they have two alleles for each gene.
- If the two alleles for a gene are the same, the individual is **homozygous**. An individual that is homozygous is **pure-breeding**. This means that, when crossed with an identical homozygous individual, all the offspring will also have the same genotype and phenotype.
- If the two alleles for a gene are different, the individual is **heterozygous**. Individuals that are heterozygous are not pure-breeding.
- The **genotype** of the individual is represented in the alleles present in its DNA.
- The **phenotype** of an individual is its physical characteristics, both its appearance and physiology (the way its cells work). The phenotype may be affected by the genes and the environment.
- A **dominant** allele is always expressed in the phenotype, when only one copy of the allele is present. A dominant allele will be expressed if homozygous or heterozygous.
- A **recessive** allele is only expressed in the phenotype when no dominant allele is present, i.e. when both copies of the allele are recessive. A recessive allele can only be expressed when homozygous.

The form and structure of DNA

DNA is a **double helix**, made of two strands that twist around each other. The two strands are held together by bases on each strand, which form a series of pairs.

The bases are:

- adenosine (A)
- thymine (T)
- cytosine (C)
- guanine (G).

A always pairs with T, and C always pairs with G.

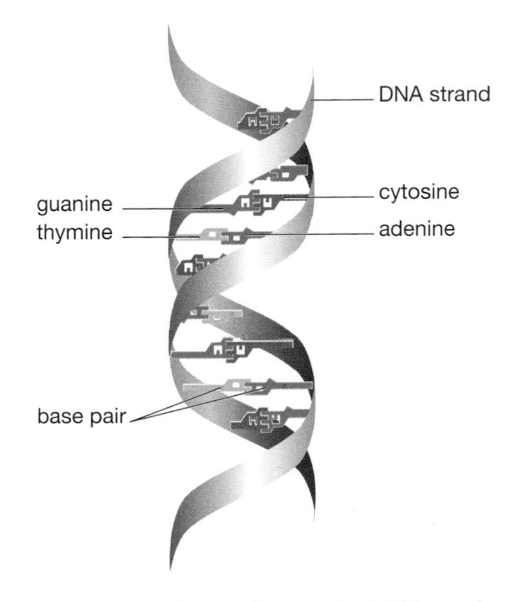

Fig. 3b.01: A short piece of a DNA molecule

Monohybrid inheritance

Monohybrid inheritance is the inheritance of a single trait or gene.

We can use **genetic diagrams** to show monohybrid inheritance. In genetic diagrams, we show genotypes and phenotypes of the parents and offspring, as well as the alleles in the gametes.

- A letter is used to represent the gene, e.g. A.
- A *capital* letter is used to show the *dominant* allele, e.g. A.
- A *small* (lower-case) letter is used to show the *recessive* allele, e.g. a.

When constructing genetic diagrams, start by stating which letter represents which allele. Choose a letter that looks different in its capital and lower-case forms.

Examples

Pea plants have a gene for flower colour. In the following diagram, F is used for the dominant allele that gives purple flowers, and f as the recessive allele that gives white flowers.

A pea plant that is heterozygous for flower colour has a genotype of Ff but a phenotype for purple flowers.

1. Starting with homozygous parents

parent phenotypes:	purple		white		
parent genotypes:	FF	x	ff		This step shows which gametes can be produced from the parents' phenotypes.
gametes:	F F		f f		
possible offspring genotypes:	Ff	Ff	Ff	Ff	This step shows all possible combinations of alleles in offspring from these gametes.
possible offspring phenotypes:	all purple				

An alternative way of showing this cross is using a **Punnett square**:

		male gametes	
		f	f
female gametes	F	Ff purple	Ff purple
	F	Ff purple	Ff purple

These four squares show the possible offspring.

2. Starting with heterozygous parents

parent phenotypes:
parent genotypes:
gametes:
possible offspring genotypes:
possible offspring phenotypes:

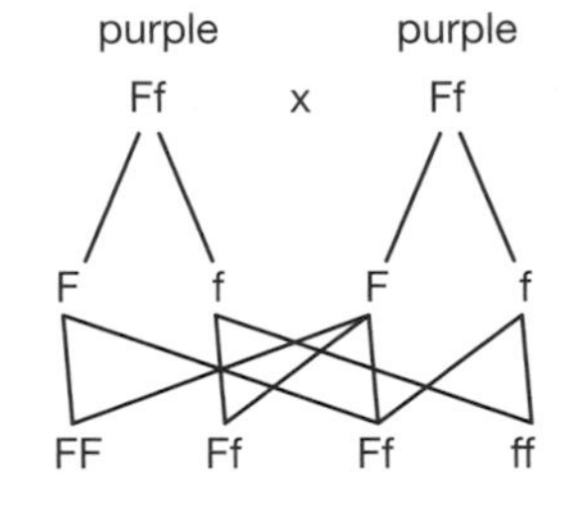

The genotype ratio of the offspring is 1 homozygous dominant : 2 heterozygous : 1 homozygous recessive.

The phenotype ratio of the offspring is 3 purple : 1 white.

Crosses of heterozygous parents produce a theoretical 3 : 1 ratio in the phenotypes of the offspring.

3. Starting with a homozygous recessive parent and a heterozygous parent

		male gametes	
		F	f
female gametes	f	Ff purple	ff white
	f	Ff purple	ff white

These crosses produce a theoretical 1 : 1 ratio in genotype and phenotypes of the offspring.

Note
Remember that genetic diagrams only give the ratio of possible offspring, not the ratio of actual offspring. Due to random fertilisation, the actual value may be very different.

Codominance

In **codominance**, different alleles inherited for a gene are both expressed in the phenotype. This occurs in the inheritance of blood group in humans. The three alleles involved are I^A, I^B and I^O.

- I^A and I^B are codominant.
- I^O is recessive to both I^A and I^B.

Possible genotypes and phenotypes are:

I^AI^A I^AI^O	blood group A
I^BI^B I^BI^O	blood group B
I^AI^B	blood group AB
I^OI^O	blood group O

A cross between a mother who is I^AI^O and a father who is I^BI^O could produce children of any blood type:

		father's gametes	
		I^B	I^O
mother's gametes	I^A	I^AI^B blood group AB	I^AI^O blood group A
	I^O	I^BI^O blood group B	I^OI^O blood group O

Family pedigrees

A **family pedigree** shows the occurrence of a characteristic in one family over several generations. They are useful because we cannot do large-scale breeding experiments on humans to understand the inheritance of traits. They can also be used to predict inheritance in a future generation.

There are standard rules for drawing and interpreting a pedigree.

- Different shapes are used to represent man and woman, e.g. a square for man, circle for woman.
- A horizontal line joining a man and woman means they are a couple who have produced children.
- Vertical lines lead to children who are listed in birth order from left to right.
- Individuals with the characteristic are differentiated from those without by shading the shapes.

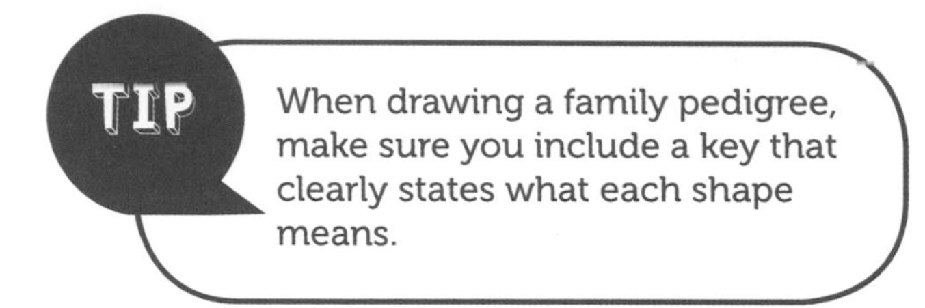

This family pedigree shows the inheritance of cleft chin in humans over three generations.

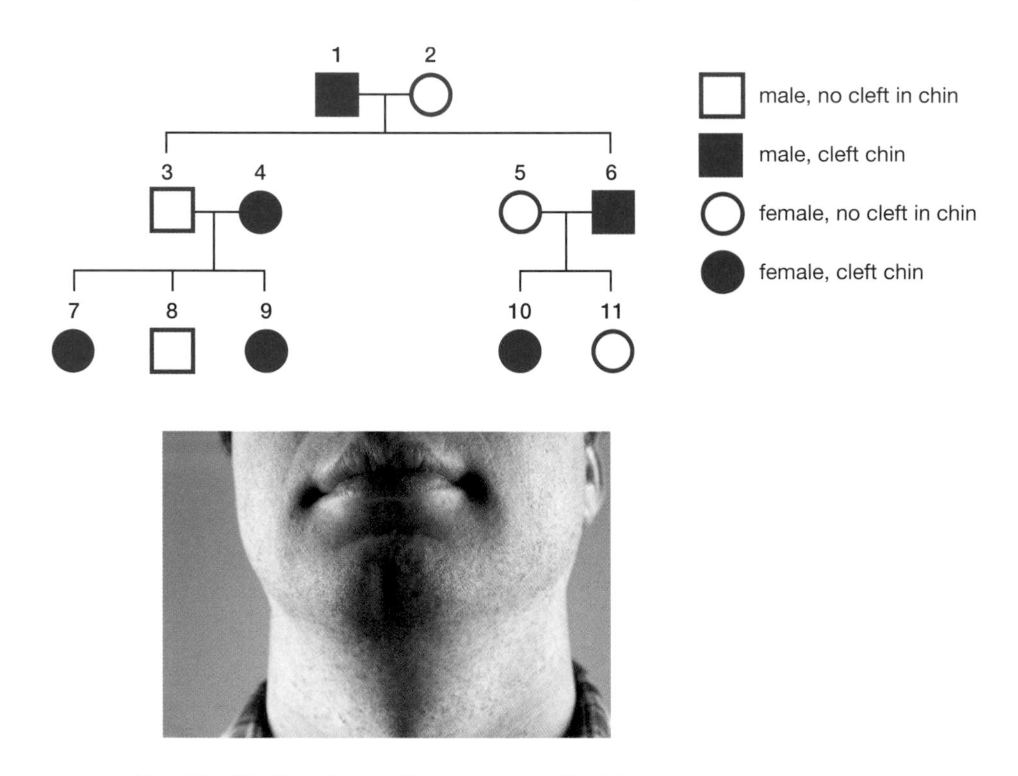

Fig. 3b.02: Family pedigree for cleft chin

From the pedigree we can see that cleft chin appears in every generation, even though the other parent in each case does not have a cleft chin. This suggests that cleft chin is dominant.

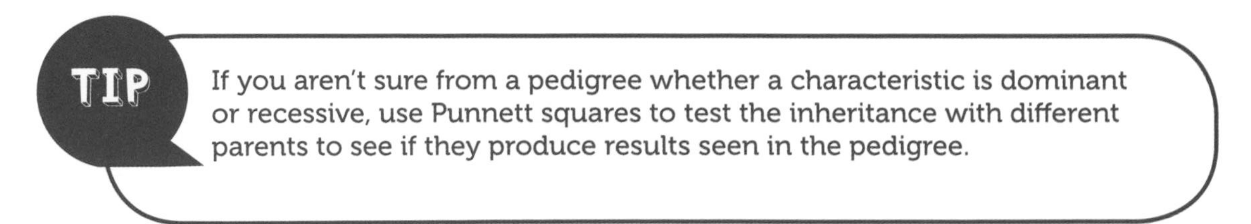

Inheritance of sex

Every diploid cell in a human contains 46 chromosomes, or 23 pairs. One of these pairs is the **sex chromosomes** as they determine the sex of a human.

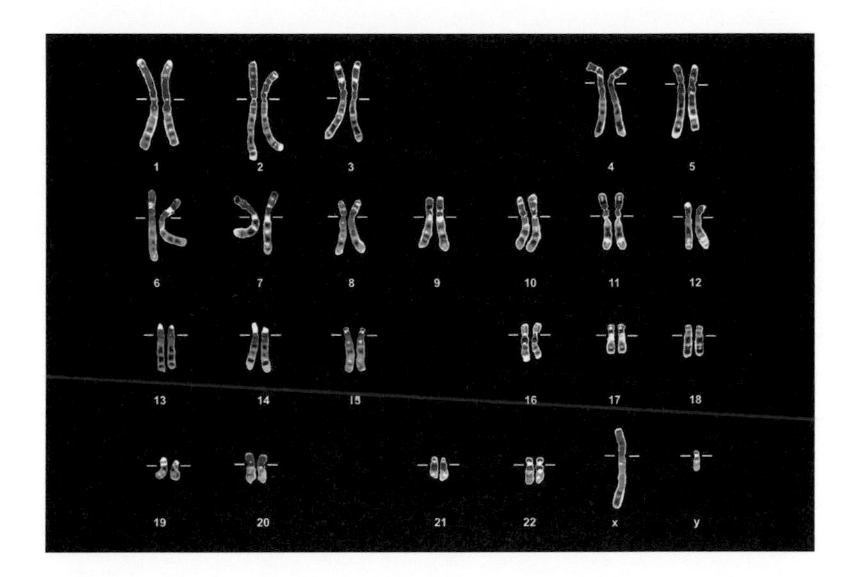

Fig. 3b.03: The chromosomes from a man's body cell, arranged in pairs. Note the difference between the X and Y chromosomes

The human sex chromosomes are called X and Y because of the way they look under the microscope. The Y chromosome is smaller than the X chromosome.

- Females have two X chromosomes: XX.
- Males have one X and one Y chromosome: XY.

Humans always inherit an X chromosome from the mother, because all her gametes contain an X chromosome.

A child can inherit an X or a Y chromosome from the father because half of his gametes contain an X chromosome, and half contain a Y chromosome.

We can show inheritance of sex using a Punnett square:

		father's gametes	
		X	Y
mother's gametes	X	XX female	XY male
	X	XX female	XY male

Note that this Punnett square shows chromosomes, not alleles

This shows that the chance of each fertilisation producing a boy or a girl is 50%, or that the theoretical ratio of boy to girl babies produced is 1 : 1.

Cell division

Cells produce other cells by cell division.

Mitosis

Mitosis is the division of a diploid cell (usually a body cell) to produce two diploid daughter cells that contain identical sets of chromosomes.

Before mitosis begins the chromosomes replicate to make identical copies of themselves. The sets of chromosomes then separate into the two new cells.

The diploid number of chromosomes is maintained in mitosis.

Mitosis produces new cells for growth, repair, cloning and asexual reproduction (see Section 3A).

Meiosis

Meiosis is the division of a diploid cell to produce four haploid cells (there are two divisions of the parent cell to make four daughter cells).

Since a copy of only one chromosome of each chromosome pair from the diploid cell ends up in each haploid cell, and there may be different alleles for each gene on the chromosomes in the pair, the haploid cells produced are not genetically identical.

In plants and humans, meiosis produces the gametes (sex cells).

When fertilisation occurs between two haploid gametes, the original diploid chromosome number is restored in the zygote.

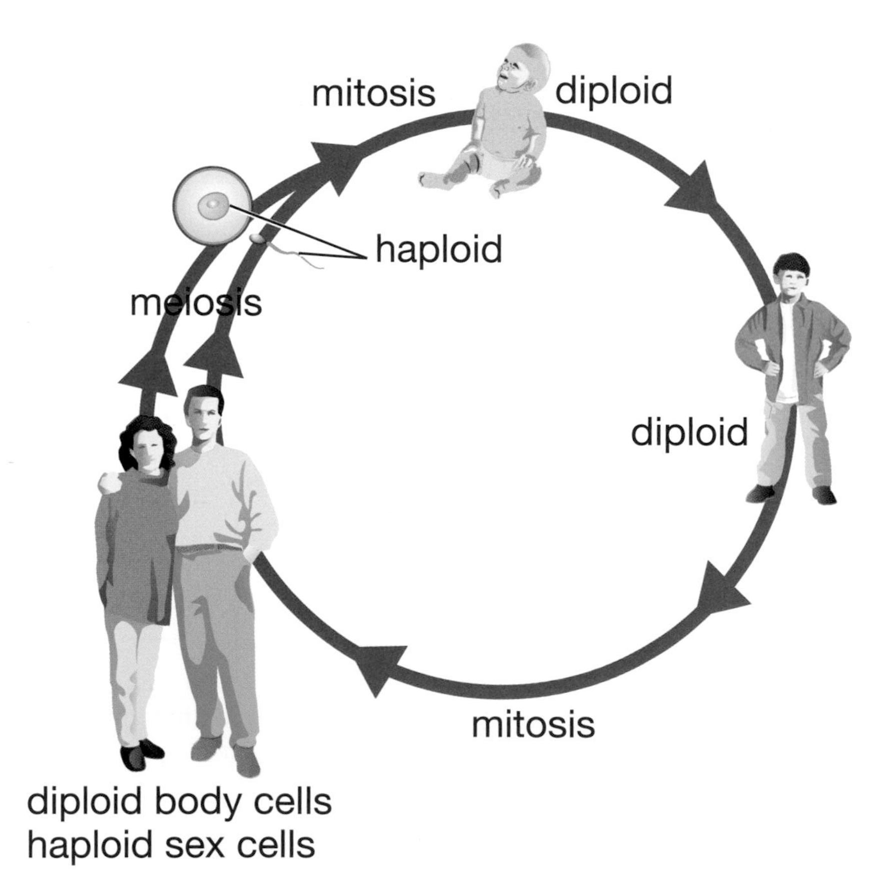

Fig. 3b.04: The human life cycle

Variation

During sexual reproduction, which sperm fuses with the egg is the result of chance (**random fertilisation**). As gametes are genetically different, this means the offspring will be genetically different from each other. This produces **genetic variation**.

Examples of genetic variation in humans include: cleft/smooth chin, free/attached ear lobes, eye colour.

During growth and development, the environment can cause variation in the phenotype that is not genetic.

Examples of **environmental variation** include: failure to grow well due to disease, lack of nutrients, lack of light (in plants).

Variation is often due to a *combination* of genetic and environmental factors. For example:

- you may have inherited genes for tallness from both your parents, but you might not grow to be tall because of illness or malnutrition in childhood
- grass turns yellow if it is covered, and turns green again when it is exposed to light.

CAM

Continuous and discontinuous variation

Continuous variation occurs when the characteristic measured varies in a continuous way, such as height in humans.

This is usually the result of a combination of genetic and environmental factors, which cause the characteristic to vary across a range in a continuous way.

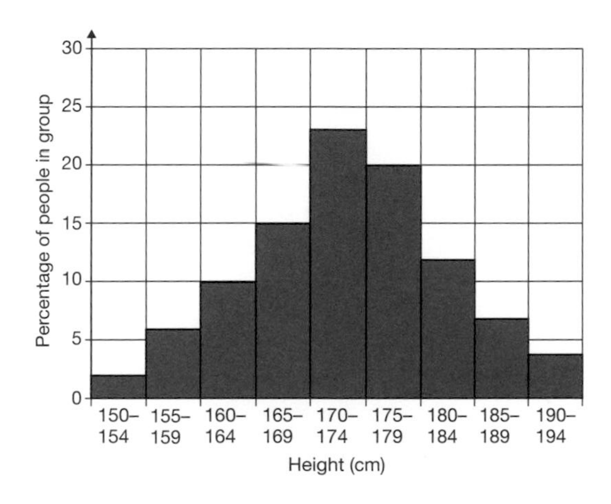

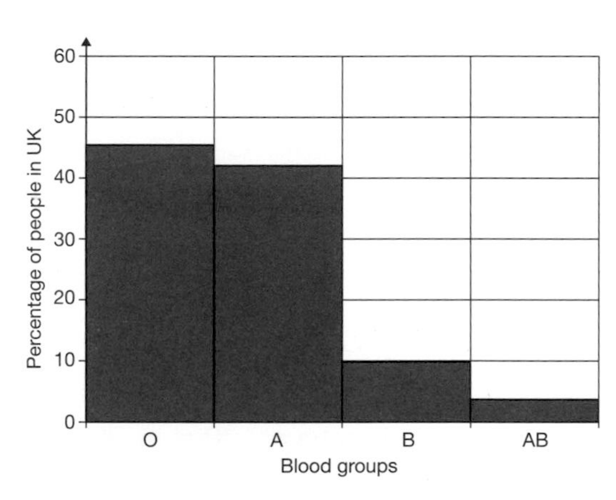

Fig. 3b.05: (Left) Continuous variation in human height, (Right) Discontinuous variation in human blood groups

Discontinuous variation occurs when the characteristic occurs in a limited number of phenotypes, such as blood group in humans. This variation is caused by genes alone.

If you aren't sure if some data are continuous or discontinuous, ask yourself if an intermediate makes sense. You can have a height of 174.6 cm but you cannot have a blood group of AO.

Mutation

A **mutation** is a change in the DNA, that is, a change in the base sequence of a gene or the gain or loss of a whole or part of a chromosome. Mutations therefore cause genetic variation.

Mutations are:

- rare events – there is a one base pair change per 10^9 bases copied each time a human cell divides
- random – they can occur anywhere in the DNA.

If the mutation occurs in a gamete, the mutation can be inherited by the offspring.

Mutations cause variation, e.g. an extra chromosome in a zygote can cause the individual to develop with Down's syndrome (an inherited condition). This occurs when a person has three copies of chromosome 21 instead of the usual two copies.

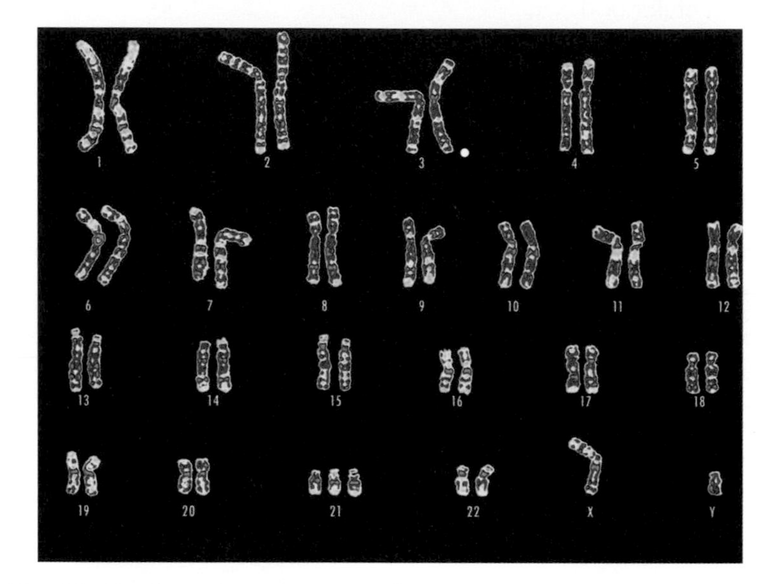

Fig. 3b.06: The chromosomes from the body cell of a person with Down's syndrome – note the extra chromosome 21

Mutations can be:

- harmful – a mutation in a cell can lead to cancer
- neutral – many mutations have no effect at all on a cell
- beneficial – a mutation can improve the chance of survival of the individual, e.g. the sickle cell allele in areas of malaria; antibiotic resistance in bacteria.

Sickle cell anaemia

A mutation in the gene for haemoglobin can produce the recessive sickle cell allele. The sickle cell allele produces haemoglobin that becomes long and thin when blood oxygen concentration is low. This causes the red blood cells to become sickle-shaped.

If an individual is homozygous for this allele, they will suffer from **sickle cell anaemia** where all the red blood cells contain sickle haemoglobin. When oxygen concentration is low, the sickled red blood cells become stuck in capillaries, which causes pain and means tissues are starved of oxygen.

If an individual is heterozygous for this allele, they have **sickle-cell trait**, which has no effect on them except under anaesthesia. However, in areas where the malaria parasite occurs (mainly tropical and subtropical wet areas), the one sickle cell allele increases the chance of survival when infected with malaria. So in these areas, the allele is beneficial.

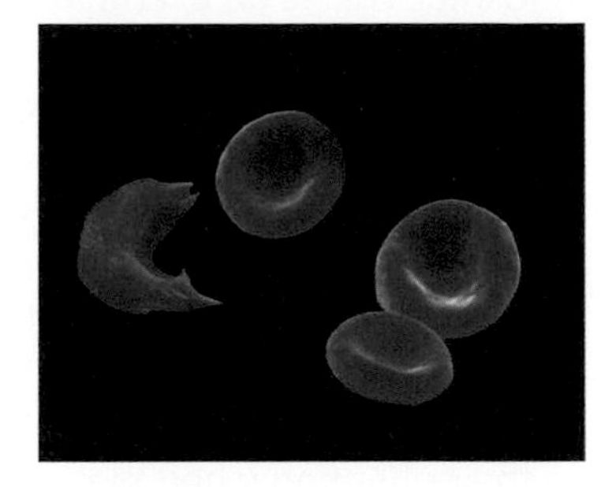

Fig. 3b.07: An electron micrograph of sickled and normal red blood cells

Evolution by natural selection

Evolution is the change in a species over time. All species of living things evolved from simple life forms (bacteria). Simple life forms first developed over three billion years ago.

Species evolve because of **natural selection**, where conditions in the environment mean that some individuals survive better and produce more offspring.

Evolution by natural selection occurs because of the following:

- In every species the individuals show variation in phenotype, some of which is due to genetic variation in the genotype.
- Some individuals are better adapted to the environment than others.
- The individuals in the population compete for food and other resources. Individuals with the best adaptations for that environment are more likely to survive and reproduce.
- The offspring inherit genes from their parents. So, if individuals with particular adaptations produce more offspring, there are likely to be more individuals with those genotypes in the next generation.
- From generation to generation, the proportion of individuals in the population with the advantageous alleles will increase over time, and the characteristics of the species will change (evolve), so much so that a new species evolves that cannot interbreed with the original one. This process is called speciation.

Natural selection leads to organisms that are better adapted to survival in a particular environment. Natural selection acts differently on two populations of a species that become isolated geographically, and can lead to speciation.

When the environment changes, however, the advantages or disadvantages of different phenotypes might change. Natural selection does not necessarily result in long term progress in a set direction. Extinction can be caused by new predators, diseases or competitors, or can also be caused by a catastrophe, e.g. volcanic eruptions or meteor strikes.

Evolution of antibiotic resistance in bacteria

Antibiotics such as penicillin are used to treat bacterial infections. They work by killing bacteria or stopping them from growing.

Antibiotic resistance in bacteria occurs when bacteria are no longer affected by an antibiotic. This is an example of natural selection.

- Bacteria in a population show variation. Some of this is genetic variation, including increased resistance to an antibiotic. This occurs naturally in a few individuals as a result of mutation.
- When an antibiotic is used to treat a bacterial infection, most bacteria will be killed. If the antibiotic treatment is not continued for long enough, the few individuals with the mutation that makes them more resistant will survive.
- The individuals with the antibiotic-resistance mutation pass the mutation on to their offspring when they reproduce (divide).
- If bacteria with the mutation infect another human, that infection will not respond as well to the antibiotic, and the person may become very ill.

Taking a full course of antibiotics is important. It reduces the number of bacteria and so helps the body's white blood cells to kill all the bacteria, including those that are more resistant naturally to the antibiotic.

Some bacterial populations, e.g. MRSA (methicillin-resistant *Staphylococcus aureus*), are resistant to many kinds of antibiotics.

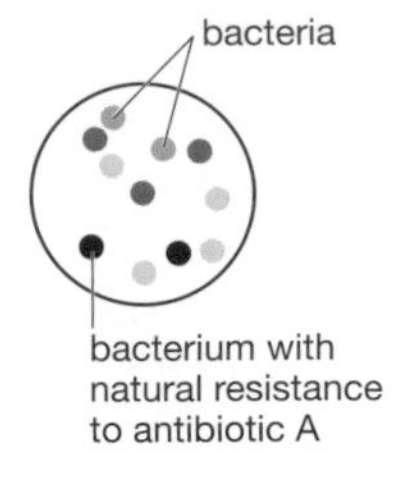

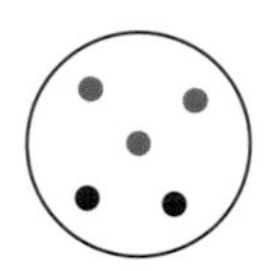

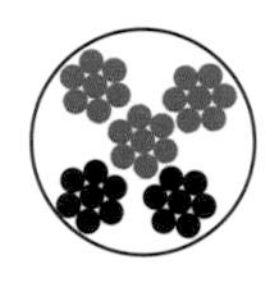

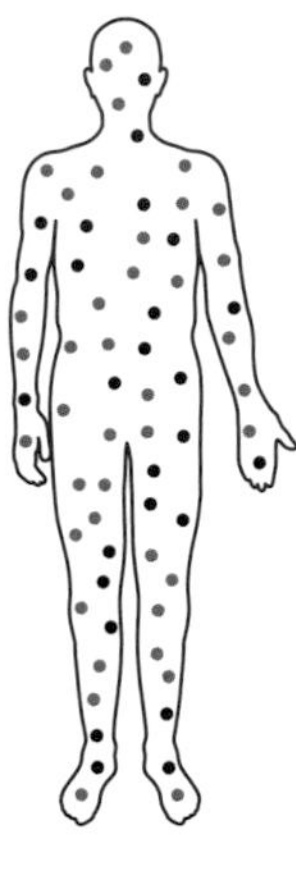

Fig. 3b.08: Stages of antibiotic resistance

Causes of mutation

Changes in DNA can be caused by:

- **ionising radiation**, e.g. gamma rays, X rays or ultraviolet rays (including from sunlight), can lead to the development of cancers (e.g. skin cancer from too much sunlight)
- **chemical mutagens** (chemicals that cause changes), e.g. the chemicals in tobacco smoke that are associated with increased risk of lung, mouth and other cancers.

You should now be able to:

★ define the term inheritance (see page 122)
★ describe the relationship between gene, nucleus, chromosome and cell (see page 122)
★ define the term *gene* (see page 122)
★ describe the structure of a DNA molecule (see page 123)
★ explain how alleles give rise to differences in inherited characteristics (see page 123)
★ explain the difference between a dominant allele and a recessive allele (see page 123)
★ define the terms *homozygous*, *heterozygous*, *phenotype* and *genotype* (see page 123)
★ give an example of codominance and explain what it means (see page 125)
★ use a genetic diagram to show inheritance in a monohybrid cross of your choice, and describe the probability of each outcome of that cross (see page 124)

★ explain how to interpret a family pedigree (see page 126)
★ describe the determination of the sex of offspring at fertilisation, using a genetic diagram (see page 127)
★ compare mitosis and meiosis (see page 128)
★ describe the sources of variation in individuals (see page 129)
CAM ★ define and give examples of continuous variation and discontinuous variation (see page 129)
★ define the term *mutation*, and describe some of its causes (see page 130)
★ describe the process of evolution by means of natural selection (see page 131)
★ describe how resistance to antibiotics can increase in bacterial populations (see page 132).

Practice questions

1. A boy has Down's Syndrome. You are given a photograph taken of the chromosomes from a body cell.

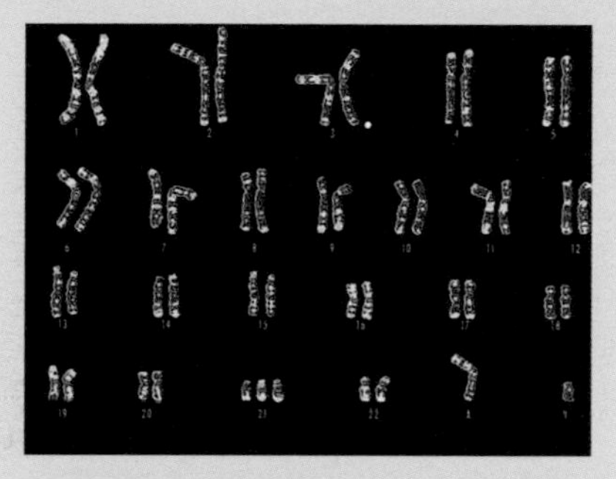

(a) How can you tell the chromosomes are from a male? **(1)**

(b) How can you tell that this boy has Down's syndrome? **(1)**

(c) Usually Down's syndrome is caused by a problem during the development of the egg. What is the problem that occurs during meiosis? What happened at fertilisation to produce the baby boy with Down's syndrome? **(2)**

(d) Huntington's is an inherited disease caused by a dominant allele. Show a simple family pedigree of a mother affected by Huntington's, a normal father and four children – two boys and two girls. One of the boys and one of the girls has Huntington's. Key: plain circle normal female, shaded circle Huntington's female; plain square normal male, shaded square Huntington's male. **(4)**

(e) Draw a Punnett square of the inheritance of Huntington's in the above family pedigree. Use H for Huntington's allele and h for normal allele. **(4)**

(f) What is a genetic disease? **(2)**

2. German cockroaches are household pests found throughout the world. In the USA, cockroach traps have been used for decades. These cockroach traps contain an insecticide mixed with glucose. The glucose attracts the cockroaches to the trap, the cockroaches eat the poisonous glucose and die. In California, a population of German cockroaches has been discovered that does not go near the cockroach traps because these cockroaches dislike glucose. Explain how this population of glucose-hating German cockroaches evolved. **(5)**

Section Four

4 Ecology and the environment

A The organism in the environment

You will be expected to:

- ★ define the terms *population*, *community*, *habitat* and *ecosystem*
- CAM ★ describe factors that affect the rate of population growth

- ★ describe the stages in a sigmoid population growth curve
- ★ describe the increase in human population size and its social implications
- ★ interpret graphs of human population growth
- ★ describe how to use quadrats to estimate population size
- ★ describe how to use quadrats to sample the distribution of organisms.

Definitions in ecology

- A **population** is all the individuals of the same species living in a particular area at the same time, e.g. a population of one species of woodlouse in a rotting log.
- A **community** is all the populations of every species living in a specific area, e.g. all the woodlice, spiders, fungi, etc. living in or on the rotting log.
- A **habitat** is the place where an organism or population lives, e.g. a rotting log.
- The **ecosystem** is all the living organisms (the community) plus all the physical factors such as amount of light, nutrients, water, temperature, availability of water, oxygen, carbon dioxide and territory (nesting sites, shelter and space to grow) of the environment in a particular area, interacting through the energy flow and cycling of nutrients.

CAM

Population growth

Population growth is the difference between the increase in new individuals (e.g. by reproduction) and the decrease by loss of individuals (e.g. by death) from a population. The rate at which a population grows depends on factors including:

- food supply – if food is limited some individuals may starve to death, and fewer young may be produced, so population growth will be slower
- predation – if predation increases, there will be fewer individuals to reproduce so population growth will be slower
- disease – this may kill individuals or slow down the rate of reproduction, so population growth will be slower
- migration – movement from a population (emigration) or to another population (immigration).

Sigmoid population growth curve

A growth curve for a population that grows rapidly in an environment with limited resources may show a sigmoid shape as in Fig. 4a.01. The curve can be divided into four stages.

The change in stage is caused by a change in conditions.

- **Lag phase**: there is no obvious increase in cell numbers because cells are making new molecules, ready for reproduction.
- **Exponential (log) phase**: reproduction occurs at the maximum rate for the specific set of growth conditions; birth rate is much faster than death rate.
- **Stationary phase**: no change in population size because death rate is equal to the birth rate, caused by a limiting factor, e.g.:
  - a required nutrient is being used up
  - waste products increase and inhibit reproduction.
- **Death phase**: birth rate is reduced by a limiting factor to the point where death rate is greater than birth rate and population size falls.

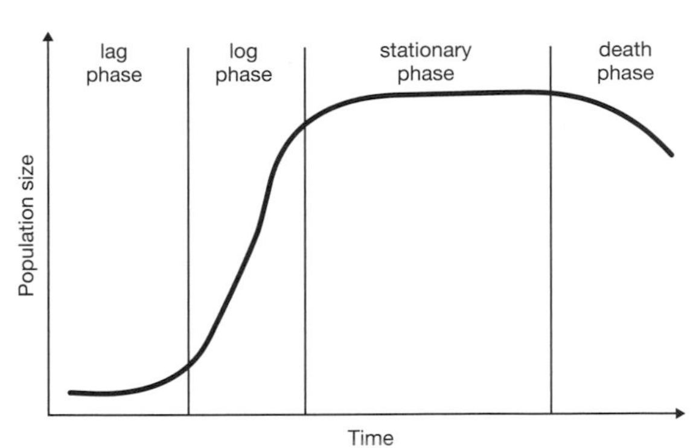

Fig. 4a.01: Sigmoid growth curve

Human population growth curve

Human population size is growing rapidly. Even if the rate of growth stays constant, population size will increase because people are living longer.

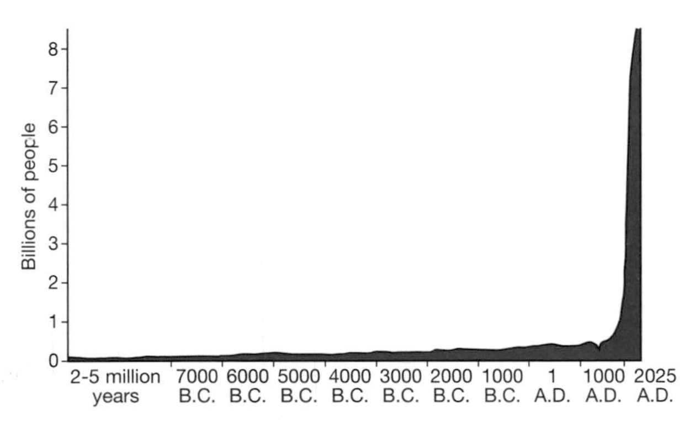

Fig. 4a.02: Human population through history

Human population has increased because:

- agriculture makes it possible to grow more food than we could acquire by hunting or foraging
- technology makes it easier to grow more food and transport it all over the world
- technology also makes it possible to change the environment (e.g. land and water supplies) to suit human needs, so more people can live together in the same place
- medicine and better hygiene reduce the human death rate.

→

The impact of increasing human population includes:

- less space for other organisms
- some resources (e.g. fresh water) at risk of becoming limited
- changing technology is altering and damaging the environment (e.g. pollution and climate change), for humans and other organisms
- increased risk of famine and war (competition for land and food).

Prediction of human population size depends on calculating changes in birth rate and death rate as a result of many factors.

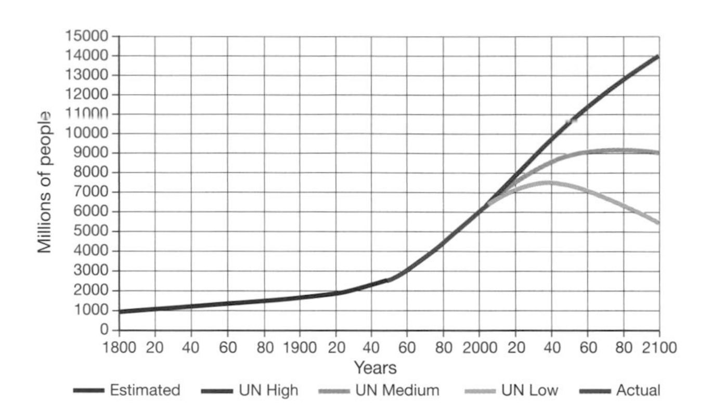

Fig. 4a.03: Human population size prediction

If birth rate is controlled, the world human population may start to fall from about 2040. If birth rate remains high, the world human population size may continue to increase for the rest of this century.

Quadrats

Quadrats are square frames of a particular size, e.g. 0.5 or 1.0 m side, used to sample the distribution of organisms in their habitats. They are best used with plants, or small animals that do not move quickly, e.g. snails.

The quantity of the plant or animal species under investigation is counted. The more quadrats used in sampling an area, the greater the sample size. This improves validity, reproducibility and repeatability.

Fig. 4a.04: A quadrat used to sample plants

Quadrats may be placed:

- randomly in a large area to estimate population size, e.g. for comparing two different areas
- systematically along a line (transect) to measure how the organisms are distributed across a changing area, e.g. from shaded to unshaded areas of a field, across a rocky shore from low tide to high tide levels.

In random sampling:

- placing of the quadrat is usually decided randomly. A calulator is used to select numbers which are used to define coordinates on a grid, centred on a corner of the sample area
- several quadrat samples are taken and the mean is calculated and used to improve the reliability of the results
- the average for one quadrat is then extrapolated to the whole area to estimate the total population size
- quadrats can also be used to estimate percentage area covered by the plant or animal under investigation.

You should now be able to:

★ explain how the following terms are related: *population*, *community*, *habitat*, *ecosystem* (see page 136)

CAM ★ describe factors that limit the rate of population growth (see page 136)

★ describe the stages in a sigmoid growth curve, and explain why they occur (see page 137)

★ describe the increase in human population size and its social implications (see pages 137–138)

★ describe how you would use a quadrat to estimate the population size of an organism in two different areas (see page 139)

★ describe how you would use a quadrat to sample the distribution of organisms in a habitat (see page 139).

Practice questions

CAM **1.**

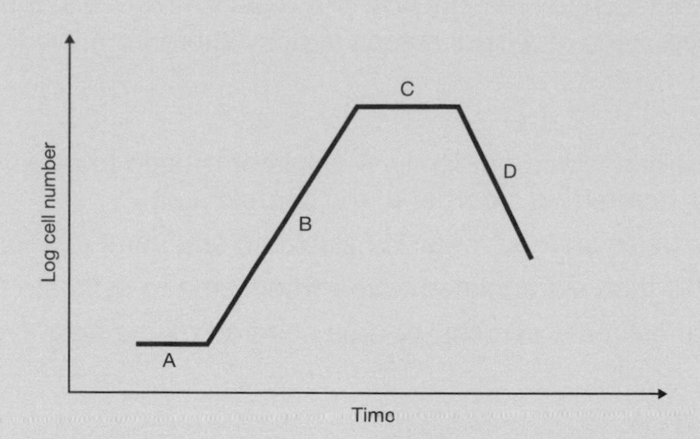

The graph above shows how the population of bacteria growing in a cold cup of tea left in your bedroom changed over four days.

Answer the following using the letters on the graph.

(a) When did the population of bacteria increase most rapidly? **(1)**

(b) When did the death rate become greater than the rate of reproduction? **(1)**

(c) When did the rate of population growth start to get affected by limiting factors? **(1)**

(d) State three factors that may limit the size of the bacterial population. **(3)**

(e) Section D on the graph ends at day 4. On a copy of the graph complete the graph to show what happens after day 5. **(2)**

(f) Why did this happen on day 5? **(1)**

2. The graph below shows the mass in tonnes of cod caught off Canada's Atlantic coast.

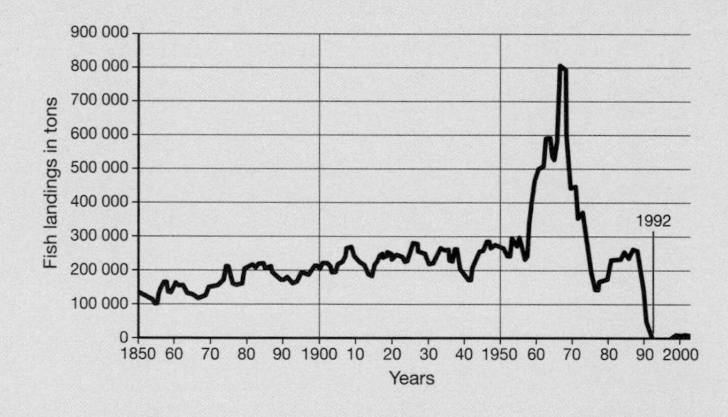

(a) Suggest and explain the reason for the serious decline in cod caught in 1992. **(2)**

(b) Canada has now banned cod fishing in the Atlantic. Suggest one ecological and one economical effect of banning fishing. **(2)**

(c) From the late 1950s to the late 1960s, there was a huge increase in the number of cod caught. Give one reason for this. **(1)**

(d) Cod eat young lobsters. The number of lobsters along the Canadian Atlantic coast has increased enormously since the late 1960s. Suggest a reason for this. **(2)**

B Feeding relationships

You will be expected to:

★ name the different trophic levels
★ describe what is represented by food chains and food webs
★ interpret pyramids of number, biomass and energy transfer
★ describe the transfer of substances and energy along a food chain
★ explain why only a small proportion of energy is transferred to the next trophic level
CAM ★ explain why food chains usually have fewer than five levels
★ explain differences in efficiency in human food chains.

Trophic levels

A **trophic level** is a feeding level in a food chain or food web, e.g. producers, primary consumers, secondary consumers, tertiary consumers and decomposers.

- A **producer** is an organism that makes its own food, e.g. plants and protoctists that photosynthesise using energy from the Sun to convert inorganic water and carbon dioxide to organic sugars.
- A **consumer** is an organism that obtains its energy from eating the tissues of other organisms.
- A **primary consumer** is an organism that obtains its energy from eating producers, also called a **herbivore** if the producers it eats are plants.
- A **secondary consumer** is an organism that obtains its energy from eating primary consumers, also called a **carnivore** as it eats meat.
- A **tertiary consumer** obtains its energy from eating secondary consumers, also a carnivore.
- A **decomposer** is an organism that gets its energy from the tissues of dead organisms or waste products of organisms. They break down the organic chemicals in dead tissues and waste products to carbon dioxide, water and mineral ions, e.g. nitrates.

Consider which organism belongs at each level when looking at food webs and food chains.

Food chains and food webs

A **food chain** shows the flow of energy or food through trophic levels.

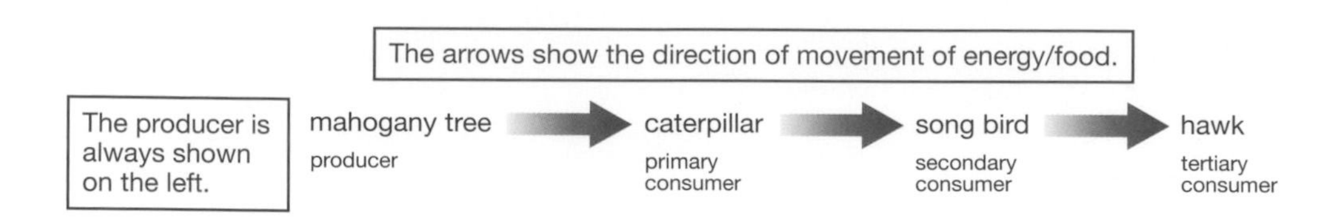

Fig. 4b.01: Food chain

A **food web** shows the interconnected food chains for many organisms in the same community.

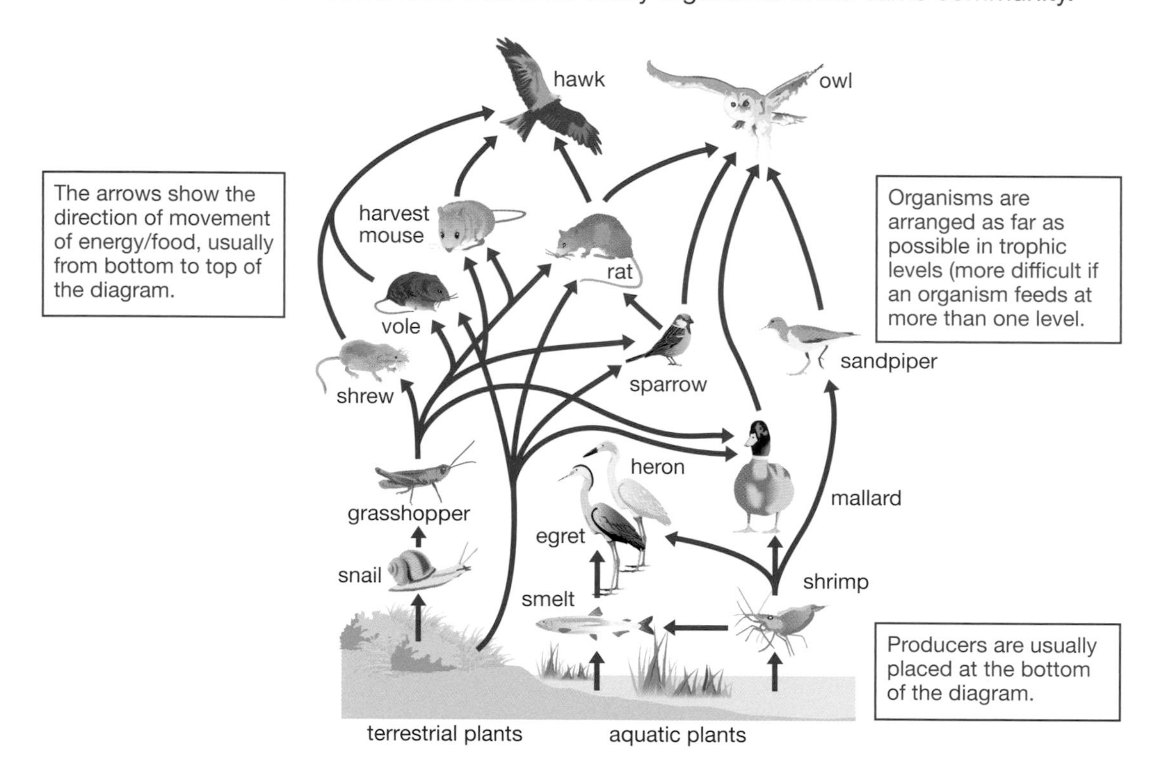

Fig. 4b.02: Food web

Fig. 4b.03: This timber wolf is a predator at the top of its food chain, where there is less energy than at the bottom; therefore there are always fewer wolves than the animals wolves prey upon; wolves have to spend most of their energy hunting their prey

Transfers through food webs

Energy transfers

The energy flow in a food chain is non-cyclical.

- Most food chains start with plants, as they start with the capture of light energy from the Sun.
- The energy is transferred to chemicals in plant tissues (chemical energy).
- The energy is taken in by primary consumers when they eat the plants: some of this energy is transferred to chemicals in animal tissue; some is lost as chemical energy in faeces and urine; the rest is lost as heat energy to the environment as a result of respiration.
- When a secondary consumer eats the primary consumer, this process is repeated, and again with higher trophic levels.
- At each trophic level, heat energy is lost to the environment, so only a small proportion of the energy gained from food is stored as chemical energy in the animal's tissues. Heat energy from the environment cannot be used for growth, etc. so this is a permanent loss of energy from organisms to the environment.

AM

- In the top trophic level, there is insufficient chemical energy in the animal tissues to sustain another trophic level (because it takes more energy to hunt and kill the food than is gained from eating the food). Most food chains are therefore limited to fewer than five trophic levels.

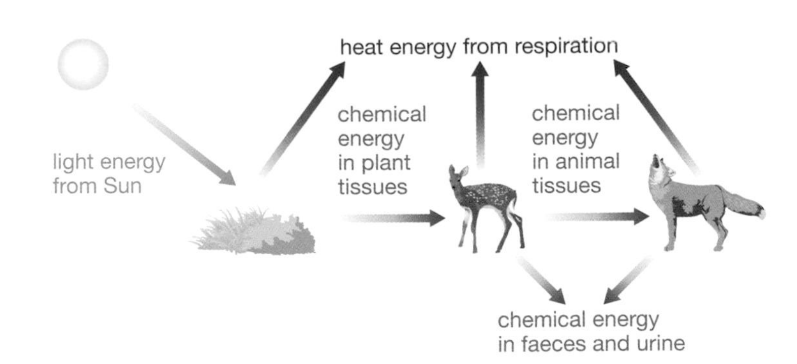

Fig. 4b.04: Energy transfer in a food chain

CAM

Efficiency in human food chains

- Humans feed as primary consumers when we eat plants, or secondary (or higher) consumers when we eat animals.
- Since energy is lost to the environment at each trophic level, there is more energy available to us if we reduce the number of stages in a food chain, i.e. if we eat the plants and not the animals that eat plants.

Transfer of substances

The transfer of substances between the environment and living organisms is cyclical. (See Section 4C for the water cycle, carbon cycle and nitrogen cycle.)

Trophic pyramids

Trophic pyramids are drawn to help us understand the transfers between organisms in a food chain.

They are drawn as a series of rectangles for each trophic level, starting with producers at the bottom. The width of the rectangles shows the factor in proportion for each trophic level.

Pyramids of number

A **pyramid of number** shows the number of organisms in each trophic level of a food web.

They may be a traditional pyramid shape, or inverted if they start with a single large producer such as a tree.

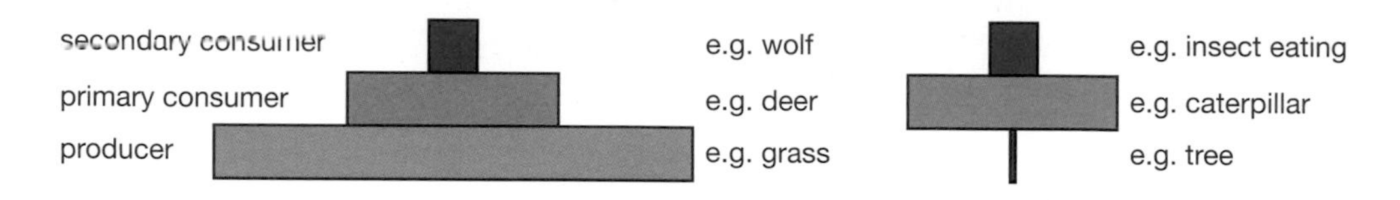

Fig. 4b.05: Two pyramids of number

Pyramids of biomass

Pyramids of biomass show the amount of **biomass** that is available at a given moment in each trophic level. This would change the pyramid for the tree food web (see Fig. 4b.05) into a pyramid shape.

Pyramids of biomass are not always pyramid-shaped because population biomass can change rapidly over time (e.g. over winter, many plants die back completely but the top consumers are still there).

Pyramids of energy

Pyramids of energy show the energy in each trophic level. Since the amount of chemical energy stored in body tissue decreases in each trophic level, these diagrams are always pyramid shaped.

You should now be able to:

★ define these terms: *trophic level*, *producer*, *consumer*, *decomposer* (see page 141)
★ explain what a food chain and a food web show (see page 142)
★ describe the conventions for setting out food chains and food webs (see page 142)
★ describe the movement of energy and substances through ecosystems (see page 143)
★ explain why only a small proportion of energy is transferred to the next trophic level (see page 143)
CAM ★ explain why food chains are limited in length (see page 143)
★ explain why, in terms of energy, it is more efficient for humans to eat plants rather than animals (see page 143)
★ describe, with examples, what pyramids of number, biomass and energy show (see above).

Practice questions

1.

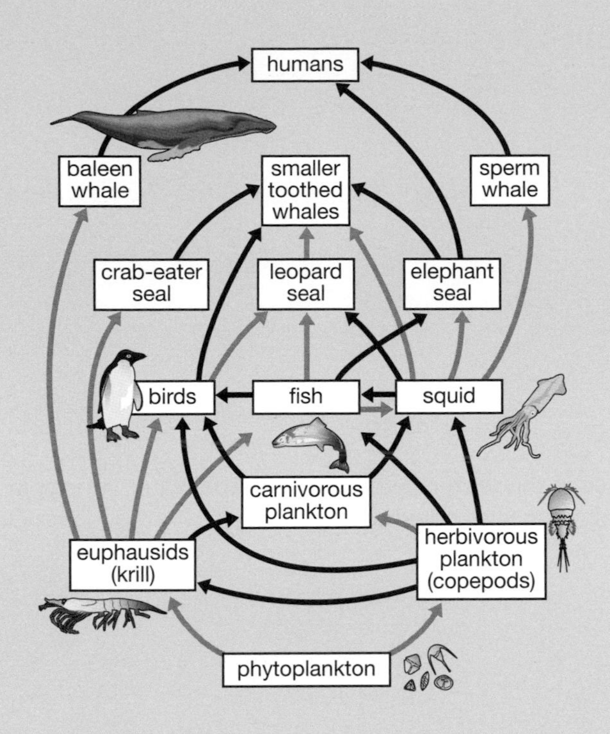

(a) The above diagram is a marine food web. Phytoplankton are producers. Name the process that occurs when energy is absorbed by the producers. **(1)**

(b) What energy change occurs during this process? **(1)**

(c) List three ways in which energy is lost from this food web. **(3)**

(d) What does each arrow represent? **(1)**

(e) Humans have reduced the numbers of baleen and sperm whales that are caught. What effect will this have on the crab-eater, leopard and elephant seal populations? **(2)**

2. Refer to the table below of a woodland community in England.

Organism	Mass of organism (kg)	Daily energy requirements (kJ)	Energy stored (kJ)
Owl	1	330.0	4000
Stoat	0.1	80.0	400
Field mouse	0.025	20.0	100
Wild oat grass	0.005	5.5	12

(a) How many stoats does an owl need to eat in a week? **(1)**

(b) How many wild oat grass plants does a mouse need to eat in a day? **(1)**

(c) How many field mice does a stoat have to eat in a day? **(1)**

(d) Draw a pyramid of numbers for the above food chain for a week. **(4)**

C Cycles within ecosystems

You will be expected to:

- ★ describe stages in the water cycle, including evaporation, transpiration, condensation and precipitation
- ★ describe stages in the carbon cycle including respiration, photosynthesis, decomposition and combustion
- ★ describe stages in the nitrogen cycle including the role of bacteria.

The water cycle

On a global scale, water circulates in the **water cycle** between the water vapour in the air, through living organisms and the water in rivers, lakes and oceans.

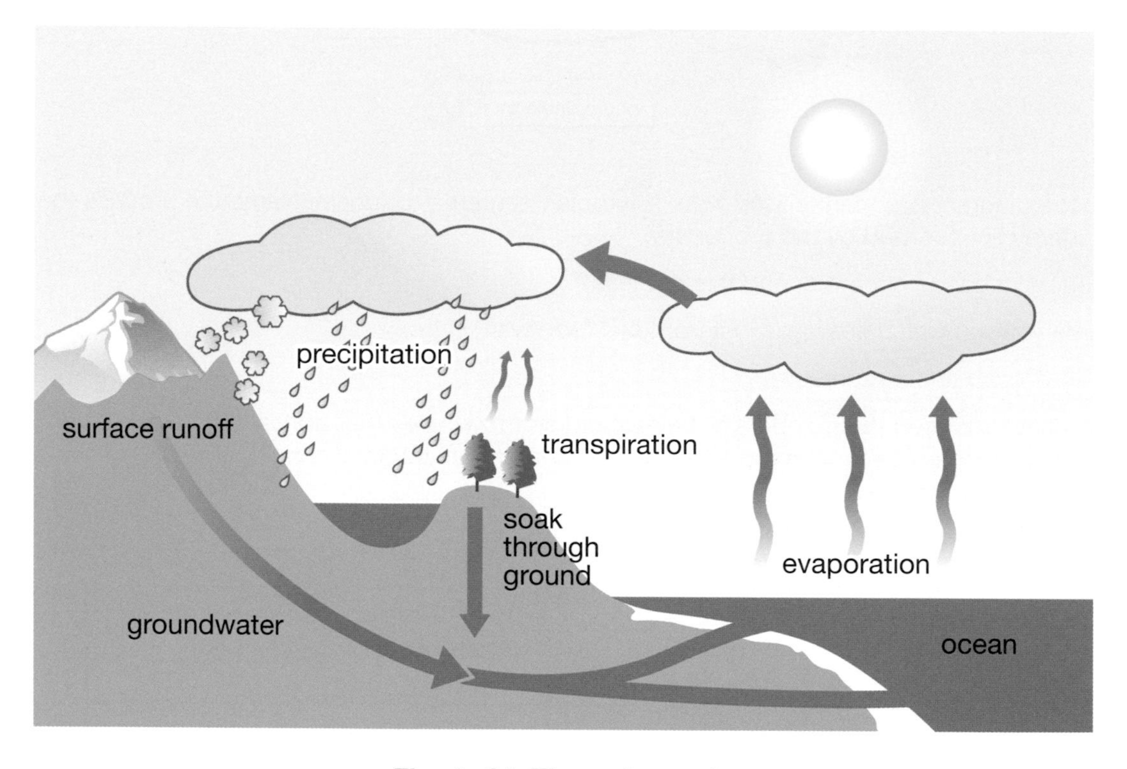

Fig. 4c.01: The water cycle

Four processes are key to the water cycle:

- **evaporation** from oceans, soils, lakes, rivers, converts liquid water to water vapour in the atmosphere
- **transpiration** from plant surfaces, converts liquid water in the plant to water vapour in the atmosphere
- **condensation** of water vapour in the atmosphere to droplets of liquid water in clouds, or even to solid ice
- **precipitation**, when the liquid water from the clouds falls as rain, or solid water falls as snow or hail. This water then returns to the oceans via rivers, coastal runoff and underground flows.

Clouds move due to the wind, so precipitation usually falls far from where the water evaporated. Most water evaporates from the oceans.

The carbon cycle

The **carbon cycle** describes the transfer of carbon between carbon dioxide in the air and organic forms of carbon in the body tissues of plants, animals, decomposers and fossil fuels.

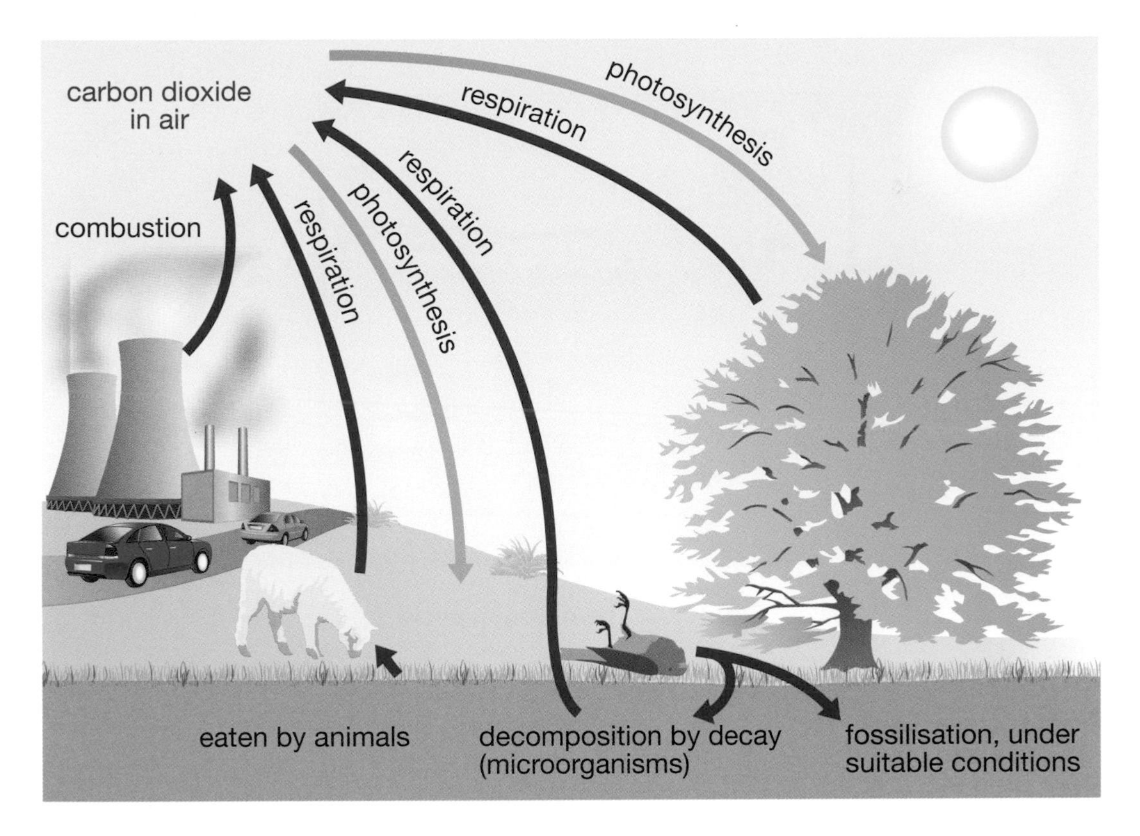

Fig. 4c.02: The carbon cycle

The key processes in the carbon cycle are:

- **photosynthesis** – converts *carbon dioxide* in the air to *organic carbon compounds* in plant tissue and in algae within oceans, ponds and lakes
- **respiration** of all organisms that releases carbon from *organic carbon compounds* as *carbon dioxide* into the air
- **decomposition** of organic materials in dead organisms and their waste by decomposers – some of the organic carbon is converted to *organic carbon compounds* in cells of the decomposers and some is released as *carbon dioxide* into the air by respiration
- **combustion** of fossil fuels – fossil fuels (e.g. coal, oil, natural gas) formed from dead organisms millions of years ago, and are therefore stores of *organic carbon compounds*. Combustion releases the carbon as *carbon dioxide* into the air.

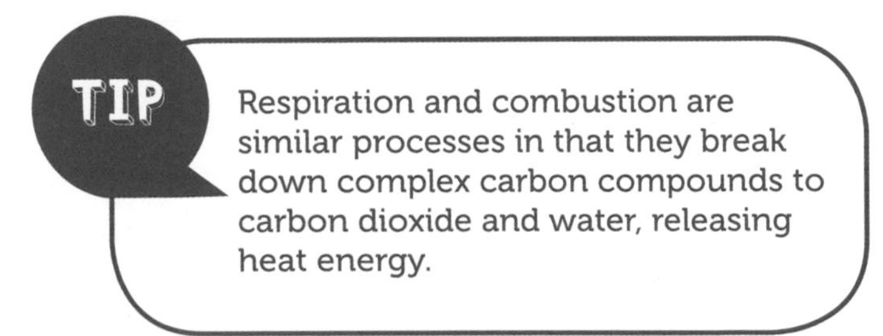

The nitrogen cycle

Organisms need nitrogen, e.g. to make proteins. Nitrogen makes up 78% of the atmosphere, but it is chemically inert and unavailable to most living organisms.

Bacteria play an important role in the nitrogen cycle.

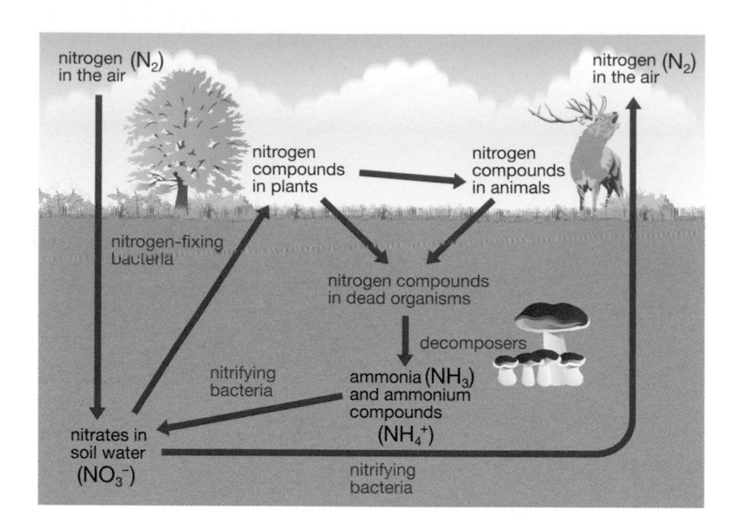

Fig. 4c.03: The nitrogen cycle

- **Nitrogen-fixing bacteria** in the soil convert nitrogen gas to nitrates – this is called **nitrogen fixation**.
- Nitrogen-fixing bacteria in the roots of some plants (legumes, e.g. peas, beans and clovers) convert nitrogen gas to nitrates inside the plants.
- Plants absorb nitrates from the soil and use them to make organic nitrogen compounds such as proteins.
- Animals get the nitrogen they need in organic nitrogen compounds from eating plants or other animals.
- Decomposers break down dead plant and animal tissues, releasing ammonia and ammonium compounds into the soil.
- **Nitrifying bacteria** convert the ammonia and ammonium compounds to nitrate ions.
- **Denitrifying bacteria** break down organic nitrogen in the soil and release it as nitrogen gas back into the air.

TIP

Remember: nitrogen-fixing bacteria and nitrifying bacteria make nitrogen available to plants.

Denitrifying bacteria reduce the amount of nitrogen available to plants.

You should now be able to:

★ describe the stages in the water cycle (see page 146)
★ describe the stages in the carbon cycle (see page 147)
★ describe the roles of nitrogen-fixing bacteria, nitrifying bacteria and denitrifying bacteria in the nitrogen cycle (see above).

Practice questions

1. A sugar refining factory pumps liquid effluent into a river. This effluent contains sugar. The diagram below shows changes in water conditions for 10 km downstream from where the sugar refining factory pours out its effluent.

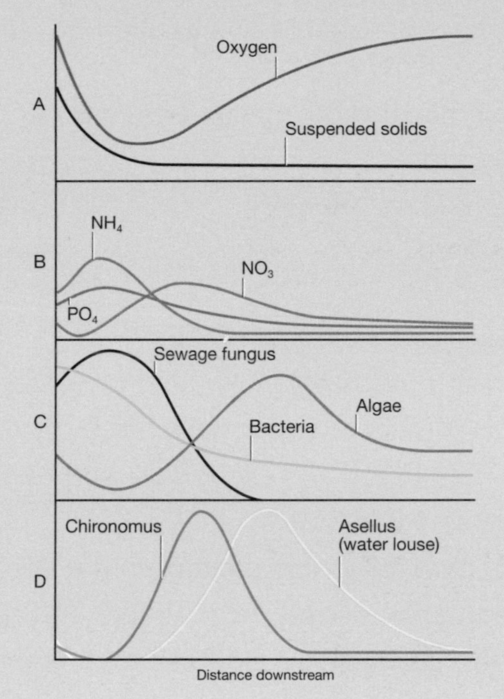

(a) Explain the shape of the curve for the quantity of bacteria downstream from the outflow. **(2)**

(b) Oxygen demand is the amount of oxygen needed by organisms living in a river. State how the oxygen demand changes as the number of bacteria in the river water increases. **(2)**

(c) Fish require a lot of oxygen for respiration. On a copy of the graph draw a line for the number of fish downstream from the outflow of effluent. **(2)**

(d) If the oxygen concentration increases with distance downstream, suggest two ways in which oxygen may enter the water. **(2)**

2. (a) In the carbon cycle, name the process by which plants absorb carbon dioxide from the atmosphere. **(2)**

(b) Name the process by which animals release carbon dioxide into the atmosphere. **(2)**

(c) How do animals obtain the carbon from plants? **(2)**

(d) Over time and under the right conditions, plants can be turned into peat or coal. **(2)**

(i) How does the carbon in coal become part of the air again? **(2)**

(ii) What environmental problem does this cause? **(2)**

(iii) What process turns dead animals into carbon dioxide? **(2)**

D Human influences on the environment

You will be expected to:

- ★ describe the effect on living organisms of air pollution by sulfur dioxide and carbon monoxide
- CAM ★ discuss the causes of acid rain and how it can be reduced
- ★ name the main greenhouse gases
- ★ describe how human activities contribute to greenhouse gases
- ★ describe the link between greenhouse gases and the enhanced greenhouse effect and global warming
- ★ describe possible consequences of global warming
- ★ explain how pollution of water by sewage can lead to depletion of oxygen in the water
- CAM ★ describe some undesirable effects of water pollution by chemical waste
- ★ describe some effects of deforestation, both locally and on a wider scale
- CAM ★ describe pollution due to nuclear fallout
- ★ discuss the effects of non-biodegradable plastics in the environment
- ★ explain the need for conservation of species, habitats and natural resources
- ★ explain how limited and non-renewable resources can be recycled.

Pollution is the addition of substances to the environment that causes harm to living organisms.

The major effects of human activity on ecosystems include:

- pollution of air
- pollution of water
- deforestation.

All of these may have local, regional and global effects.

CAM

TIP Research examples of air pollution by international companies to use in your exam answers.

Sulfur dioxide

Sulfur dioxide is an acidic gas, produced when fossil fuels (e.g. coal, oil) burn. The main source of sulfur dioxide in the air in countries with large industries is from combustion during industrial processes.

Sulfur dioxide reacts with the water droplets in clouds to form sulfuric acid. This can fall as **acid rain**, sometimes many kilometres from the source of the gas (e.g. pollution from the UK can damage ecosystems in Sweden).

Sulfur dioxide in air can dissolve in moisture on breathing surfaces, e.g. in the lungs, causing additional difficulty for people with breathing problems.

Acid rain makes soil, lakes and streams more acidic. This can:

- damage plant roots, causing plants to die
- react with soil to release toxic chemicals
- damage organisms that live in water and so alter aquatic ecosystems.

Fig. 4d.01: Damage done as a result of acid rain dissolving a limestone statue

CAM

Reducing acid rain

The amount of acid rain produced can be controlled by reducing the emissions of sulfur dioxide. In many industrialised countries, emissions have fallen greatly over the past few decades because of laws relating to the following:

- coal used for burning at home must have most of the sulfur removed
- gases from coal and oil burnt in power stations and industry must be cleaned of sulfur dioxide before they escape to the air
- petrol and diesel used in transport must have most of the sulfur removed before use.

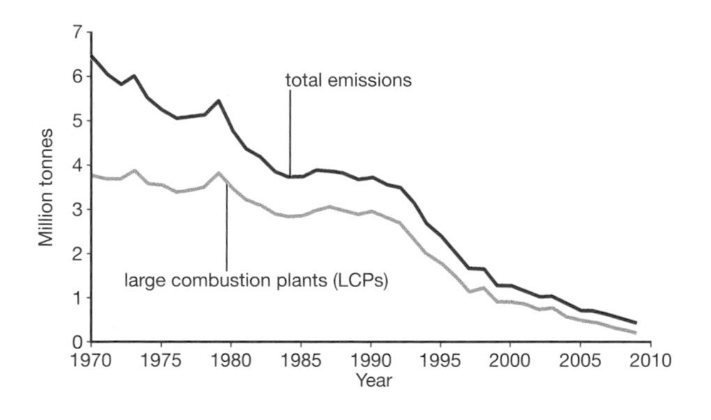

Fig. 4d.02: Sulfur dioxide emissions in the UK

Carbon monoxide

Carbon monoxide is a toxic gas. It forms a stable compound with haemoglobin, preventing red blood cells from carrying oxygen. This can stress the heart, reduce oxygen flow to tissues and is particularly harmful to rapidly developing tissues, e.g. in babies and young children.

Carbon monoxide is produced by the incomplete combustion of carbon-containing compounds. In the air it comes mostly from transport and industry. Use of catalytic converters on cars has greatly reduced carbon monoxide emissions.

Greenhouse gases

Greenhouse gases trap heat energy radiated from the Earth's surface, as a result of the **greenhouse effect**. This is a natural process and keeps the Earth's surface warm, which makes it possible for living organisms to survive on the surface. CFCs (chlorofluorocarbons) have a different mechanism. CFCs combine with the ozone high up in our atmosphere, and destroy it. The ozone layer protects Earth by absorbing much of the UV radiation from the Sun. CFCs mean more UV radiation now reaches Earth.

The main greenhouse gases and their sources are:

Gas	Natural sources	Human activities that release gas
water vapour	evaporation transpiration	combustion
carbon dioxide	respiration	combustion of trees, fossil fuels and peat
nitrous oxide	nitrogen cycle (denitrifying bacteria)	combustion
methane	natural decay in waterlogged conditions, and in guts of herbivores	farming, e.g. keeping cows, growing rice
CFCs (chlorofluorocarbons) destroy the ozone layer	none	were used as coolants, e.g. in refrigerators, as solvents, in plastic foams, and as propellants in aerosol cans – being phased out

Increases in greenhouse gases in the air can cause the **enhanced greenhouse effect** by trapping even more heat in the atmosphere. This may be the cause of **global warming**, the increase in temperature of the Earth's surface over the past few hundred years.

Global warming could lead to **climate change** and other problems by causing:

- increased desertification in the centres of continents
- more floods and storms along coastal areas
- rising sea levels due to ice melt and sea water expansion
- spread of tropical contagious diseases to cooler regions
- changes in distribution of species
- changes in animal migration patterns and plant flowering times.

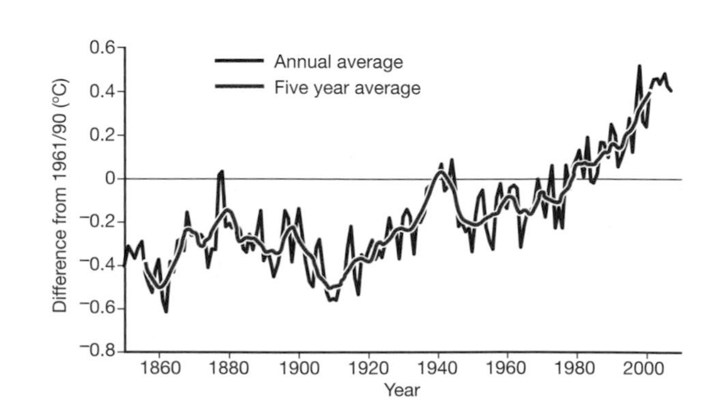

Fig. 4d.03: Measurements of the temperature of the Earth's atmosphere since 1850 shows global warming

Water pollution

Sewage contains human waste and detergents making it nutrient-rich, especially in phosphates and nitrates.

Fertilisers containing high levels of minerals, e.g. phosphates and nitrates, are used on farmland to improve crop growth.

Water pollution can occur when:

- sewage gets into water systems
- excess fertilisers leach through the soil and get into streams, rivers and lakes.

Water pollution can also occur when chemicals from industrial processes get into water systems. If the chemicals are toxic, they will kill living organisms and make the water dangerous for drinking.

Increasing the concentration of minerals in water systems is called **eutrophication**. Eutrophication can lead to a loss of biodiversity (as organisms die out).

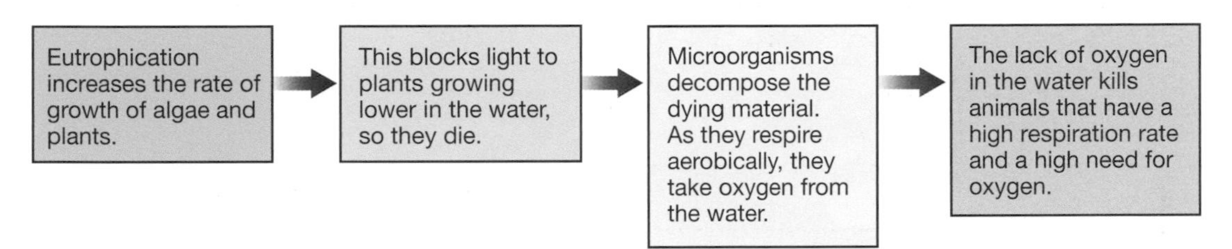

Fig. 4d.04: Effects of eutrophication

Land Use

Deforestation

Deforestation is the cutting down of trees on a large scale, such as logging of tropical hardwoods in Indonesia for timber, clearance of the Amazon rainforest for cattle ranching, mining, or for crops from which biofuel (ethanol) can be produced.

Deforestation has the following effects:

- increased **leaching**, i.e. the loss of nutrients from the soil by draining away in water which the trees would have absorbed, leaving the soil less fertile
- **erosion** of soil – tree roots help to stabilise the soil after heavy rainfall
- disturbance of the water cycle – trees absorb water from rainfall and transpire it into the atmosphere, reducing the amount of water that is lost in surface runoff that causes flooding and erosion
- disruption of the carbon cycle – trees act as long-term carbon stores. The carbon from the cycle is lost quickly in the burning of forests – less carbon dioxide is removed from the air when the trees have been cut down and less oxygen is added by photosynthesis
- habitat destruction – loss of biodiversity
- increase in methane – deforested areas are used to raise cattle and grow rice.

Fig. 4d.05: Soil erosion/flooding after deforestation

Other land uses

- Humans build, farm and dump waste on land, and this use of land means other organisms lose their habitats.

Non-biodegradable plastics

Non-biodegradable plastics may take hundreds of years to decay, so they have long-lasting effects on the environment:

- nylon nets catch and kill fish, marine birds and marine mammals long after they have been used for fishing from vessels
- plastic can-loops strangle birds
- some aquatic animals swallow plastic bags, confusing them with their jellyfish prey
- when plastics decay, they may release chemicals that are toxic.

Non-biodegradable plastics pollute every region on Earth.

Fig. 4d.06: Plastic pollution of the ocean

Pollution from nuclear fallout

Radioactive material from nuclear power stations is normally tightly controlled so that it does not escape.

During a nuclear explosion, such as at Chernobyl in 1986, radioactive material may escape into the environment. The radioactivity may affect all living organisms in the area in which the material falls out of the air.

For example, after the Chernobyl explosion, the rate of thyroid abnormalities in children was seven times as high in children downwind of Chernobyl as in the area upwind. Hundreds of millions of pounds of crops and livestock were lost due to the fallout.

On 14 March 2011, a tsunami hit Japan and caused a meltdown at the Fukushima Dai-Ichi nuclear power plant. Direct radiation exposure of Fukushima residents was not as high as was initially feared, but soils across north-eastern Japan are contaminated and could affect public health for decades through the produce farmed in these areas.

Conservation

Conservation is the preservation of species, habitats and the environment. This is important because we need:

- resources, e.g. fresh water and non-renewable materials such as fossil fuels and minerals
- living resources for food, and products such as timber
- habitats that protect the environment by:
  - controlling, e.g. flood protection provided by mangroves, reefs and wetlands
  - capturing, e.g. carbon stores provided by forests
- places for cultural purposes, e.g. recreation, education, aesthetic and spiritual contemplation.

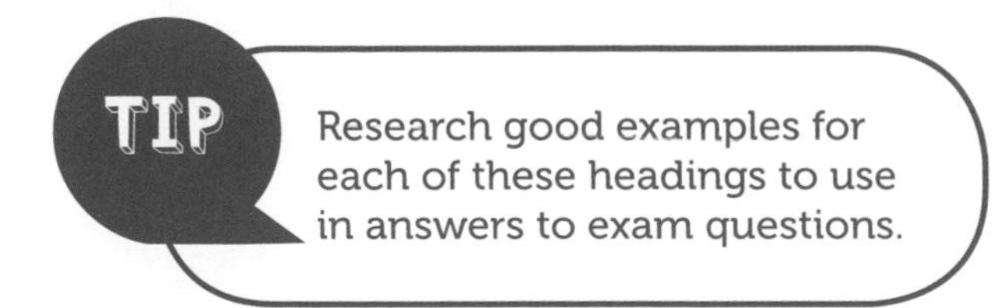

Recycling

Recycling is the recovery of material that has been disposed of, so that it can be treated in order to make it useful again. This is important because it saves depleting limited or non-renewable materials.

Paper recycling

Paper is made from wood pulp, which is a renewable resource. However huge quantities are constantly required, e.g. in the USA, every week 500 000 trees are used to make all the Sunday papers. Recycling paper:

- saves space in landfill sites
- reduces the area of land used to grow quick-growing trees for paper
- uses 30–50% less energy than making new paper
- reduces air pollution from paper manufacture by 95%.

Sewage treatment

Sewage is treated in water treatment plants by:

- sedimentation, to remove bulky matter
- decay organisms, to remove high levels of nutrients and toxic chemicals
- aeration, to increase the oxygen concentration of the water.

This makes the water safe to return to the environment without causing water pollution.

Further treatment of the water, including chemicals such as chlorine that kill microorganisms, make the water safe to drink.

Fig. 4d.07: Water treatment plant

You should now be able to:

- ★ explain how sulfur dioxide can cause pollution (see page 151)

- ★ describe how sulfur pollution of the air can be reduced (see page 151)
- ★ describe some of the effects of air pollution by carbon monoxide (see page 152)
- ★ name the main greenhouse gases (see page 152)
- ★ describe how human activities contribute to greenhouse gases (see page 152)
- ★ explain how the enhanced greenhouse effect may lead to global warming and climate change (see page 152)
- ★ explain how pollution of water by sewage causes a depletion of oxygen in the water (see page 153)
- ★ give two sources of eutrophication (see page 153)
- ★ describe the main effects of deforestation on a large scale (see page 153)

- ★ describe the need for conservation of species, habitats and natural resources (see page 154)
- ★ describe the advantages of recycling paper (see page 155)
- ★ describe the effects of non-biodegradable plastics in the environment (see page 154)
- ★ explain how sewage can be treated to produce water that is safe to return to the environment or for human use (see page 155).

Practice questions

1. In the last 150 years, it is estimated that 45% of all tropical rainforests have been destroyed, mainly by burning them.

(a) Give three reasons why this large scale deforestation occurred. **(3)**

(b) Deforestation on this scale is affecting the carbon dioxide concentrations in the air.

(i) What effect is deforestation having on the carbon dioxide levels? **(1)**

(ii) How does deforestation cause these changes? **(2)**

(iii) Explain what effect deforestation has on the climate and on soil fertility. **(2)**

(c) Describe two methods that scientists and governments use to protect the disappearing habitats and wildlife. **(2)**

2. When a gardener makes a compost heap from leaves and weeds, she covers it in a plastic sheet and once a week she mixes her compost with a garden fork. She noticed, one year, that her compost was ready earlier after a spring that was wetter than previous years.

(a) Explain why this happened. **(1)**

(b) How does mixing the compost up each week help speed up the decay? **(2)**

(c) The plastic sheet keeps in the warmth of the compost heap. Explain why a compost heap becomes warm. **(2)**

(d) Why is the compost so good for her garden plants? **(3)**

Section Five

5 Use of biological resources

A Food production

You will be expected to:

CAM ★ discuss problems of world food supplies and the problems that contribute to famine
★ discuss ways in which modern technology has increased food production
★ describe how glasshouses and polytunnels can increase crop yield
★ explain the effects of increased carbon dioxide and temperature on crops in glasshouses
CAM ★ describe the use of optimum light in glasshouses
★ explain why fertilisers increase crop yield
★ explain why pests of crop plants are controlled
★ describe the advantages and disadvantages of using pesticides and biological control
★ describe the role of yeast in making beer
★ describe a simple experiment to investigate carbon dioxide production by yeast
★ describe the role of *Lactobacillus* in making yoghurt
CAM ★ describe the use of microorganisms for manufacturing penicillin and single cell protein
★ describe and explain the features of an industrial fermenter for growing microorganisms
★ describe and explain the methods used in fish farming to increase yield.

CAM

World food supplies

Global food production has increased greatly over the past 50 years or so.

It has increased because of developing technologies:

- modern agricultural machinery, such as tractors and harvesters, makes it easier to cultivate larger areas of land (compared with using people and animals)
- chemical fertilisers, pesticides and herbicides increase crop production (see below)
- artificial selection (selective breeding, see Section 5B) improves the varieties and breeds of organisms we use so they produce more food.

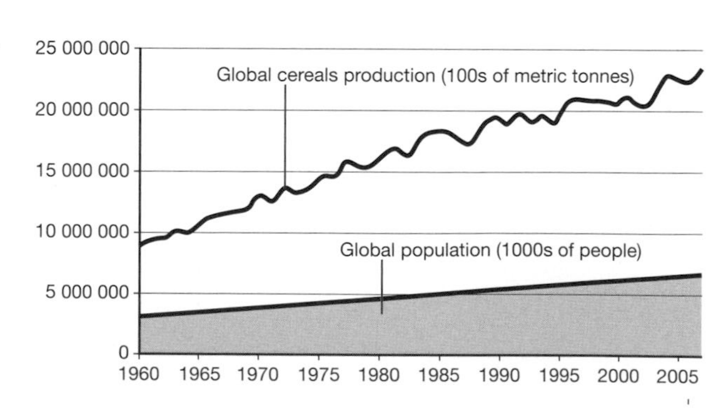

Fig. 5a.01: World food production

The global human population is increasing also, but not as fast as food production. Despite this, over 1 billion people in the world do not have enough food of the right kinds to maintain their health. This is because:

- food is not grown and distributed equally – people in richer countries generally have access to much more food than those living in poorer countries
- people living in very dry areas suffer when drought kills crops, and those living in very wet areas lose crops to flooding
- conflict, including war, displaces people from where they grow their crops, to camps where they depend on food being given to them.

Improving crop yield

The **yield** of a crop is the amount of food harvested from the crop, e.g. the weight of grain produced from a wheat crop. The rate of photosynthesis of a crop is one factor which determines the yield of the crop.

Glasshouses and polytunnels

Glasshouses (greenhouses) are metal-framed buildings with panes of glass for roofs and walls, in which plants are grown. Polytunnels are made of a metal frame covered in polyethylene plastic. Polytunnels are easier to move around than greenhouses, and are cheaper.

Glasshouses and polytunnels help with growing crops by:

- protecting the crops from extremes of weather
- keeping the soil warmer, so seeds germinate faster and crops grow more quickly
- extending the growing season so crops can be grown for longer periods
- protecting the crops from many insect pests – and any pests are usually easier to control in the restricted space.

Fig. 5a.02: Crops being grown to maximise yield in a polytunnel

The conditions in glasshouses can be controlled quite carefully to maximise growth rate.

- Additional heating can be provided by oil burners when the temperature of the surrounding environment is low, to provide the optimum temperature for enzymes and for the rate at which cell processes occur.
- If the temperature rises too high, ventilation can be provided to keep the temperature at an optimum for growth.
- Additional lighting can be provided when natural light levels are low, so photosynthesis continues for longer.
- Combustion from oil burners also supplies increased concentration of carbon dioxide in the air around the plants which increases the rate of photosynthesis (carbon dioxide can be a limiting factor).

Using fertilisers

Fertilisers contain mineral ions that plants need for healthy growth, e.g. nitrates and phosphates.

As each crop grows, it absorbs mineral ions in the soil to make new plant tissue, and therefore the amount available decreases. If more mineral ions were not added to the soil, yield would decrease with each crop. Adding fertilisers to the soil is therefore important to maintain crop yields.

There is a limit to how many mineral ions the crop plants can absorb, so any excess minerals are washed or soak away and risk causing eutrophication (see Section 4D).

Pest control

Pest control is important for a high crop yield because pests damage or kill crop plants, and so reduce yield. Pests include insects, fungi and weeds.

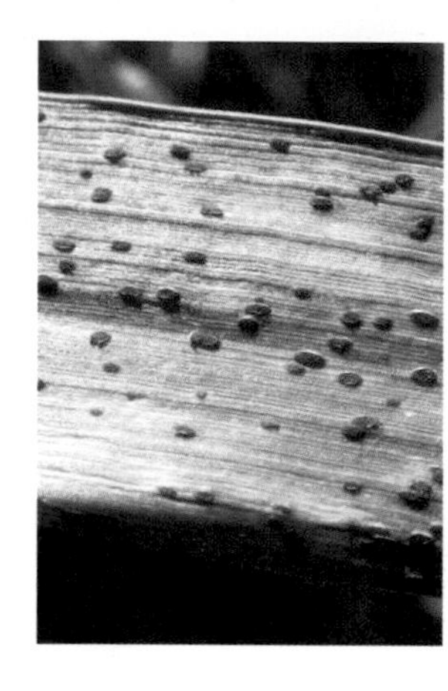

Fig. 5a.03: (Left) Aphids suck the sap from plants, so the plant has less food to convert to plant tissue (Centre) Fungal infections damage leaves, reducing photosynthesis and therefore the amount of food the plant can make (Right) Weeding is an important process to reduce competition for water and soil nutrients

When one type of crop plant is grown in a large field (**monoculture**), insect pests are more likely to be attracted and cause a lot of damage.

We use **pesticides** to kill pests. These include:

- **insecticides** to kill insects
- **fungicides** to kill fungi
- **herbicides** to kill weeds.

Many of the pesticides we use are chemicals sprayed onto the crops. For killing animal pests, we can also use **biological control**, e.g. introduce a predator of the insect, such as lacewings and ladybirds that eat aphids.

Type of pest control	Advantages	Disadvantages
chemical pesticides	• reliable, effect • can be used when needed • applied directly to the crops	• many made from non-renewable oil • stable chemicals may last a long time in the environment • may affect species other than the pest, damaging food webs • may cause health problems for farmers and other local people • insects may develop resistance to pesticide
biological control	• no chemicals used so usually less damage to environment • pest population reduced to non-damaging level, so local food webs not as affected • very useful in restricted places such as glasshouses	• can disrupt the local ecosystem if introduced species numbers increase too far and have no natural predators • organisms used as biological control may migrate away from crop • needs to be applied at the right time, when there are plenty of pests but before crop damage is a problem

Food production from microorganisms

We use many species of microorganisms to make food, including when we make cheese, bread, alcoholic drinks, yoghurt, vinegar and vegetable pickles.

Brewing beer

Yeast is a single-celled fungus that respires anaerobically in a process called alcoholic fermentation:

$$\text{sugar} \rightarrow \text{ethanol} + \text{carbon dioxide} \ (+ \text{energy})$$

Beer is an alcoholic drink made from barley grains using yeast.

- Barley grains are soaked in water to start germination, releasing enzymes that break down some of the starch in the grains to the soluble sugars maltose and glucose.
- The germinating grains are mashed and mixed with water, and heated to the optimum temperature for these enzymes. The sugars dissolve in the water.
- The barley **mash** is filtered to produce a sugary solution called **wort**.
- The wort is heated to denature the enzymes and to kill any microorganisms.
- Yeast (*Saccharomyces cerevisiae*) is added to the sugar solution. (Hops may also be added for flavour.)
- The yeast respires anaerobically, producing carbon dioxide (that gives beer its fizz) and a concentration of ethanol up to about 4%, at which concentration the yeast cells die.

Experiments with yeast

Fig. 5a.04 shows a simple apparatus for experiments with yeast. As the yeast respires, carbon dioxide is produced, and bubbles through the lime water, turning it milky.

With this apparatus you can time how long it takes for the lime water to turn milky under different conditions of: temperature, pH and sugar concentration.

Fig. 5a.04: Apparatus for yeast experiments

Yoghurt production

Yoghurt is made from milk using the bacterium *Lactobacillus bulgaricus*. This bacterium respires anaerobically, breaking down milk sugars and producing lactic acid:

$$\text{milk sugar} \rightarrow \text{lactic acid (+ energy)}$$

To make yoghurt:

- the milk is heated briefly to a high temperature (c. 70 °C) to kill off any microorganisms (pasteurisation)
- the milk is then cooled to about 40 °C and the *Lactobacillus* culture is added
- after several hours the milk is converted to a thickened yoghurt, with a slightly acidic taste caused by the lactic acid.

CAM

Single cell protein

Single cell protein is a protein produced from a fungus. The fungus is grown in large quantities in a vessel called a **fermenter** (see Fig. 5a.05) where conditions are controlled to maximise the rate of growth of the fungus.

The fungus forms hyphae, which are thread-like structures (see Section 1B). The hyphae are filtered from the growth mixture, and heat-treated to remove bitter-tasting substances. They are then pressed and dried to form a protein-rich, fibrous food that can be used as a replacement for meat.

The advantages of growing single cell protein include:

- the fermenters can be built anywhere that food is needed
- the nutrients for growing the fungus can use waste substances from other processes, such as waste straw from crop plants
- protein is produced more quickly and cheaply than growing animals, e.g. beef cattle
- single cell protein contains more fibre than meat and little fat, making it a healthier food.

Making penicillin

Penicillin is a chemical produced by the fungus *Penicillium* when growing in optimum conditions. Its effect was first investigated by Alexander Fleming when he noticed that a *Penicillium* mould was killing colonies of bacteria growing on the same agar plate.

The chemical that was killing the bacteria was extracted and purified, and called penicillin. Methods were then developed to produce penicillin in large quantities, for use as the first **antibiotic** to treat bacterial diseases.

Penicillin-like antibiotics are still very important in controlling bacterial infections.

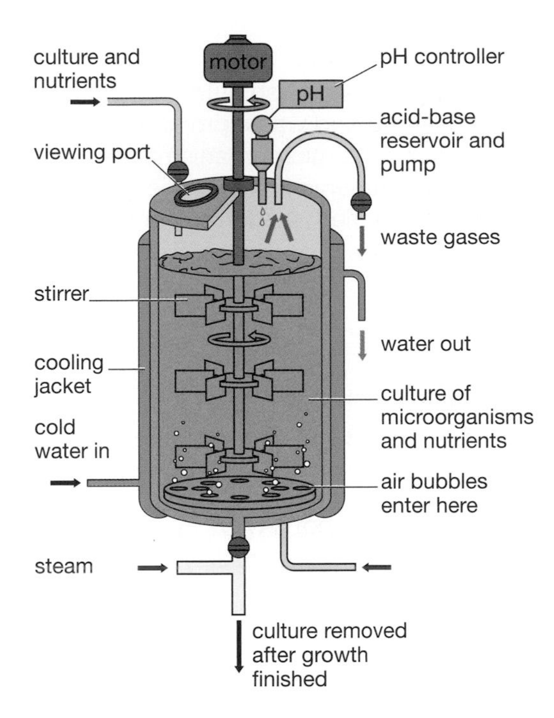

Fig. 5a.05: Industrial fermenter for growing microorganisms

Industrial fermenters

Industrial fermenters are extremely large tanks in which large quantities of microorganisms can be cultured in ideal conditions.

Before the culture is placed in the fermenter, aseptic precautions are used to prevent contamination by other microorganisms:

- the equipment is cleaned with steam
- nutrient solutions and air supplies are sterilised.

A starter culture of microorganisms is added to the fermenter with the correct balance of nutrients (e.g. amino acids, vitamins and minerals) for rapid growth.

Temperature and pH in the fermenter are kept at an optimum. Heat energy produced by respiration is removed by cool water in a surrounding cooling jacket.

To give the microorganisms aerobic conditions, a steady oxygen supply is added, and the mixture is constantly agitated (stirred) to ensure all microorganisms are in the ideal conditions for growth.

The culture is removed from the fermenter and the product from the microorganism is extracted. Fermenters are used to produce the antibiotic penicillin, as well as many other products.

CAM

Enzymes for use in biological washing products (see Section 2C), and the antibiotic penicillin are produced in industrial quantities from microorganisms grown in fermenters.

Batch culture

This system is when: a fermenter is filled with fresh nutrient medium and is inoculated with a microorganism; the nutrients are used up and waste products accumulate in the fermenter as the microorganism multiplies exponentially; the microorganisms reach a stationary phase and then death occurs when there are no more nutrients left and when waste product levels become toxic; the fermenter is emptied and the product purified; the fermenter is sterilised before another batch can be started.

Batch culture has several advantages:

- it is easy to set up
- if contamination occurs, only one batch is lost
- it is the only way to obtain certain products, e.g. penicillin is only produced by *Penicillium* mould AFTER the stationary phase.

The disadvantages are that the fermenter is not always in use as it needs to be sterilised between batches, and because the microorganisms are not always multiplying exponentially.

Mycoprotein is made by batch culture. The fermenter has a growth medium that contains glucose for respiration and ammonia as the source of nitrogen. The fermenter is inoculated with spores of the fungus *Fusarium venenatum*, and oxygen is bubbled through the medium as the fungus respires aerobically. When the fungus stops growing, the fermenter is emptied and the mycoprotein is left behind as a spongy solid mass. The mycoprotein product familiar in Europe is called Quorn. Mycoprotein, because it is from a fungus, contains roughage and has no cholesterol.

Continuous culture

This system uses the fermenter with a continuous supply of nutrients and with waste products continuously removed.

The microorganisms continue to multiply exponentially, and the desired product is produced continuously.

The main advantage is that the production rate is higher as the microorganisms are always in the exponential phase and the fermenter is kept in use for longer.

Two disadvantages are that it is more complicated to set up a continuous culture, and if contamination occurs, then huge volumes of product may have to be discarded.

Industrial ethanol is produced by continuous culture (compared to batch fermentation of wine in wine casks).

Fish farming

Overfishing of wild fish has reduced wild fish stocks resulting in not enough sexually mature fish left to reproduce. This has led to restrictions on net size, fishing quotas and the development of fish farming techniques in enclosures such as in tanks or ponds, or in cages in rivers or the sea.

To maximise growth, the fish need:

- clean water and the removal of waste products – in enclosed tanks or ponds, the water must be continually purified and replaced as the fish produce solid waste, and soluble ammonia or urea. Waste products and water may be used to grow plants such as herbs. In sea cages, the waste products are lost to the environment where they may cause problems by increasing the growth of algae and other organisms.
- frequent supplies of high protein food – in sea cages, this increases the nutrification of the surrounding environment. In enclosed tanks, this increases the need for water purification. The food may be supplied from wild fish stocks, which can increase overfishing effects.

Keeping the fish in such high densities increases the rate of disease transmission. As fish rub against each other they damage their scales, fins and tails, increasing the risk of disease. Antibiotics are used to control infections, and in sea cages these chemicals can escape into the local environment and affect other species.

Intraspecific predation (cannibalism) is the consumption of one species by the same species. Many of the fish species grown in fish farming are natural predators, e.g. salmon. In the wild, the adults would not normally live in the same environment as the young, so intraspecific predation is naturally controlled. In fish farming, the young must be kept separately from the adults of these species.

Interspecific predation is the consumption of one species by another species. In fish farming, predator species must be kept separate from herbivorous species, although different herbivorous species can be kept in the same tank or cage.

Selective breeding (see Section 5B) can be used to improve the quality of fish, e.g. for disease resistance, high reproductive rate, rapid growth rate. Fish reproduce through external fertilisation so, during the breeding season, selected females are caught and their eggs extracted by gentle squeezing. The eggs are then mixed with sperm from selected males for fertilisation.

You should now be able to:

CAM
- ★ describe some of the problems of world food supplies (see pages 158–159)
- ★ discuss ways in which modern technology has increased food production (see pages 159–160)
- ★ describe how glasshouses and polytunnels can be used to increase the yield of some crops (see page 159)
- ★ explain why crop yield in glasshouses increases as a result of increased carbon dioxide and raised temperature (see page 159)
- ★ describe the use of optimum light in glasshouses (see page 159)
- ★ explain why fertilisers are needed to increase crop yield (see page 160)
- ★ explain why pests of crop plants are controlled (see page 160)
- ★ compare the advantages and disadvantages of using pesticides and biological control with crop plants (see page 161)
- ★ describe the role of yeast in the production of beer (see page 161)
- ★ describe a simple experiment to investigate the effect of temperature on the production of carbon dioxide by yeast (see page 162)
- ★ describe the role of the *Lactobacillus* bacterium in making yoghurt (see page 162)

- ★ describe how microorganisms are used to produce penicillin (see pages 162–163)
- ★ describe how single cell protein is produced (see page 162)
- ★ describe the key features of an industrial fermenter for growing microorganisms (see page 163)
- ★ explain the methods that are used in fish farming to maximise the growth rate of fish (see page 164 and above).

Practice questions

1. Cane toads are huge toads from South America that eat all kinds of insects. In Australia, sugar cane was infested with an insect called *Pyrilla*. Cane toads were imported to Australia in the 1950s as biological pest control for the *Pyrilla* insect infestation. Unfortunately, cane toads are nocturnal and the *Pyrilla* insects are active during the day. The cane toads started to eat any small animal they could find, including many of Australia's rare small mammals. Cane toads have no natural predators in Australia.

(a) What happened to the cane toad population numbers? **(1)**

(b) What happened to the numbers of small mammals in sugar cane growing regions? **(1)**

(c) This is an example of some of the disadvantages of biological control. List three advantages of biological control. **(3)**

(d) Farmers now use chemical insecticides on the sugar cane to control the *Pyrilla* insect. Give two advantages and disadvantages of chemical pest control. **(4)**

2. Mycoprotein is made from a mould called *Fusarium*.

(a) What kingdom do moulds belong to? **(1)**

(b) Just like penicillin, mycoprotein is made in huge fermenters. What do the following structures of a fermenter do:

(i) diffuser **(1)**

(ii) water-cooled jacket **(1)**

(iii) pH probe? **(1)**

CAM (c) Penicillin is produced by batch cultivation, and mycoprotein by continuous cultivation. At the end of each kind of cultivation, the fermenter is sterilised. Why? **(1)**

(d) What is the mould called that produces penicillin? **(1)**

(e) List three advantages mycoprotein has compared to animal protein. **(3)**

B Selective breeding

You will be expected to:

★ describe how plants with required characteristics can be developed by selective breeding

★ describe how animals with required characteristics can be developed by selective breeding.

Selective breeding (artificial selection) is the selection by humans of individual plants and animals with particular characteristics for breeding.

- Breeders decide what inheritable characteristics they want to develop.
- They select individuals that show this desired characteristic most strongly to breed/reproduce with each other.
- They repeat this over many generations until they have a number of individuals with the desired characteristic.

TIP

Selective breeding is **not** natural selection because the breeders select individuals artificially.

Selective breeding provides evidence for evolution by natural selection because, over generations, characteristics of a population can be changed by selecting for them.

Selective breeding of plants

Plants are selectively bred for economic reasons such as:

- increased crop yield
- improved disease resistance
- improved resistance to environmental conditions such as frost, high temperature, drought
- improved quality of crop (sweetness, size, colour, oil content, starch content, protein content, etc.).

For example, selective breeding of wild cabbage (*Brassica oleracea*) has resulted in many different kinds of vegetables of which we eat different parts.

Some of these are listed on the following page.

Fig. 5b.01: A variety of *Brassica oleracea*

Variety of *Brassica oleracea*	Grown for
kale	large leaves
broccoli	large green stems and flowers
cauliflower	very large, white flower clusters
cabbage	very large terminal buds
brussels sprouts	large lateral buds
kohlrabi	very thick stem

In your exam, you may be given an example of selective breeding (artificial selection). You will be expected to explain which characteristic has been selectively bred for and why it is useful.

Selective breeding of animals

Animals are selectively bred for economic reasons such as:

- increased yield, e.g. milk, honey, meat, wool, eggs
- quality of product, e.g. fat content, texture, taste, colour
- increased growth rate
- improved disease resistance
- tameness
- ability to grow in harsh conditions
- reproductive rate (e.g. sheep that produce twin lambs more frequently).

For example, chickens were domesticated from wild jungle fowl in Asia. Through artificial selection we have different breeds of chicken:

- some lay a large number of eggs
- some lay large eggs
- some produce eggs with different shell colours (which are more marketable)
- some are fast-growing chickens that develop muscles (meat) quickly.

You should now be able to:

★ define the term *selective breeding* (*artificial selection*) (see page 167)

★ describe some characteristics in plants that have been developed by selective breeding (see pages 167–168)

★ describe some characteristics in animals that have been developed by selective breeding (see above).

Practice questions

1. Malaria is a disease that kills millions of people every year. It is caused by a single-celled parasite that lives in humans and in female *Anopheles* mosquitoes. The mosquitoes do not become sick, but carry the parasite which a female mosquito gets from an infected human. The parasite takes three weeks to develop in the mosquito. The female mosquito then injects the parasite into another human victim, and spreads the disease. Scientists want to breed *Anopheles* mosquitoes that only live for two weeks.

   (a) Is *Anopheles* the species or genus name of the malaria parasite-carrying mosquito? **(1)**

   (b) What is the name of the breeding process that the scientists will use to develop the *Anopheles* mosquitoes with a shorter life cycle? **(1)**

   (c) Describe this breeding process. **(4)**

   (d) Why do scientists want to develop mosquitoes with a shorter life cycle? **(2)**

   (e) Why do they not develop *Anopheles* mosquitoes that have no wings? **(2)**

   (f) What other characteristic might they want to have in the *Anopheles* mosquitoes? **(2)**

2. In the mountains of Switzerland, the winters are cold, long and very snowy. The spring and summer are short. At these altitudes, the air is thin – there is not as much oxygen as at lower altitudes.

   (a) List the characteristics farmers in the Swiss mountains wanted to have in their dairy cows. **(4)**

   (b) What did the cows need oxygen for? **(1)**

   (c) Milk prices dropped, and the farmers wanted to breed beef cows (cattle) from their herds of dairy cows. List two different characteristics they would be looking for. **(2)**

   (d) Unfortunately a new disease was introduced from herds of cows across the mountains in France, and almost all their beef cattle died. Suggest a reason why this happened. **(2)**

C Genetic modification

You will be expected to:

- ★ describe the use of restriction enzymes and ligase enzymes with DNA
- ★ describe how plasmids and viruses act as vectors
- ★ define the term *transgenic*
- ★ describe the use of genetically modified bacteria to make human insulin
- ★ evaluate the potential for use of genetically modified plants to improve food production.

Genetic modification is a process in which a gene taken from one organism is put into another organism, forming a **transgenic** organism or GM organism. The organisms may be from different species.

The transgenic organism will produce the characteristic of the inserted gene. This is possible because DNA is used in the same way in all living organisms to produce characteristics.

Enzymes in genetic modification

Restriction enzymes cut the double stranded DNA molecule at specific sites. A specific restriction enzyme is chosen to cut out the chosen gene from the DNA.

Some of these enzymes leave a short length of single stranded bases called a 'sticky end'. If the same enzyme is used on another piece of DNA, the sticky ends match up and can be joined together more easily.

Ligase enzymes join pieces of DNA together. They are used to join the chosen gene to other DNA to form **recombinant DNA**.

Vectors

The chosen gene must be inserted into the other organism. This is done using a **vector**. Vectors are pieces of DNA that naturally insert themselves into other cells. They include:

- **plasmids** – small circles of DNA found in bacteria (not the main circular chromosome of bacteria)
- viruses – forms that are non-infective are used.

The vector DNA is cut with the same restriction enzyme, and mixed with the chosen gene and a ligase enzyme, to form recombinant DNA. The vector with the chosen gene attached is inserted into the other organism.

If the other organism is a bacterium, the vector used is a plasmid, e.g. in the formation of a transgenic bacterium that produces human insulin. The plasmid with the insulin gene is copied each time the transgenic bacterium divides, making more and more bacteria that produce human insulin.

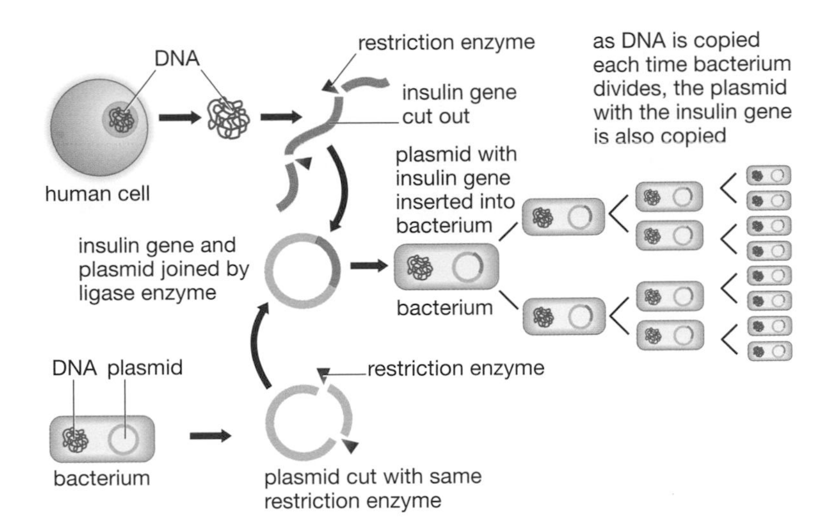

Fig. 5c.01: Formation of a transgenic bacterium that produces human insulin

If the other organism is a plant or animal, the vector must insert its DNA and the chosen gene into the DNA of the cell nucleus, so that the normal cell mechanisms copy the chosen gene and produce the required characteristic.

Industrial production of human insulin

To produce human insulin on an industrial scale, genetically modified bacteria with the human insulin gene are grown in ideal conditions in a sterilised fermenter (see Section 5A).

The bacteria reproduce very rapidly, and produce large quantities of human insulin that can be extracted and purified from the contents of the fermenter.

This makes large quantities of pure human insulin at a relatively cheap cost, for use by people who need insulin (diabetics).

Genetically modified plants

Plants can be genetically modified by giving them a gene that makes a natural insecticide. When an insect eats the cells of a plant, the cells release the toxin and kill the insect.

Plants that have been genetically modified with this gene have advantages and disadvantages.

Advantages include:

- less damage is done to the plant, so crop yield is greater
- fewer chemicals are sprayed on the plants, so less damage is done to the environment
- only insects that eat the plant are affected, which is better for the environment
- it is quicker to produce insect-resistant plants this way than by selective breeding.

Disadvantages include:

- seed for genetically modified crops can cost much more than for non-GM varieties
- crop yield may not be greater because the gene may have been inserted in a part of a chromosome that affects how well the plant grows
- insects may develop a resistance to the toxin, which means the GM variety may become useless for this feature
- effects on wild populations of plants and insects are unknown
- effects of eating GM crops on human health are unknown.

You should now be able to:

- ★ describe the action of restriction enzymes and ligase enzymes on DNA (see page 170)
- ★ explain what is meant by a *vector* (see pages 170–171)
- ★ describe how genetically modified bacteria containing the human insulin gene have been produced (see page 171)
- ★ describe advantages and disadvantages of plants that are genetically modified for insect resistance (see page 171).

Practice questions

1. Bacteria can be genetically engineered to produce human hormones such as insulin.

(a) What kingdom do bacteria belong to? **(1)**

(b) How are bacterial cells different to an animal cell? **(3)**

(c) What are plasmids? **(2)**

(d) Explain how the human insulin gene is inserted into a bacterium. **(4)**

(e) What sort of reproduction do the genetically engineered bacteria undergo to produce genetic clones? **(1)**

(f) What human disease requires the injection of insulin? **(1)**

(g) Why is insulin from genetically engineered bacteria thought to be safer than insulin extracted from other animals? **(2)**

2. Plants can be genetically engineered using viruses to insert a gene into the plant cell.

(a) What plant cell organelle does the inserted gene need to enter? **(1)**

(b) What structure in the organelle does the DNA have to be inserted into? **(1)**

(c) Genes can be inserted that improve nitrogen uptake. This means fewer fertilisers need to be used. How does this benefit the environment? **(4)**

D Cloning

You will be expected to:

★ describe the process of micropropagation of plants
★ explain why micropropagation is used to produce commercial quantities of plants
★ describe the stages in the production of cloned mammals such as Dolly the sheep
★ evaluate the potential for use of cloned mammals.

Micropropagation

Micropropagation (or tissue culture) is like taking cuttings, but at a cellular level.

- Small groups of cells, called **explants**, are taken from the growing tips of plants.
- The explants are sterilised to kill any microorganisms.
- The explants are then grown on a sterile nutrient medium, where the cells divide by mitosis to form an undifferentiated lump of cells called a **callus**.
- Cells from the callus are placed on sterile nutrient medium containing plant hormones, to cause differentiation of cells to form tiny plants with roots, stem and leaves. These are then planted out when they are large enough to handle.

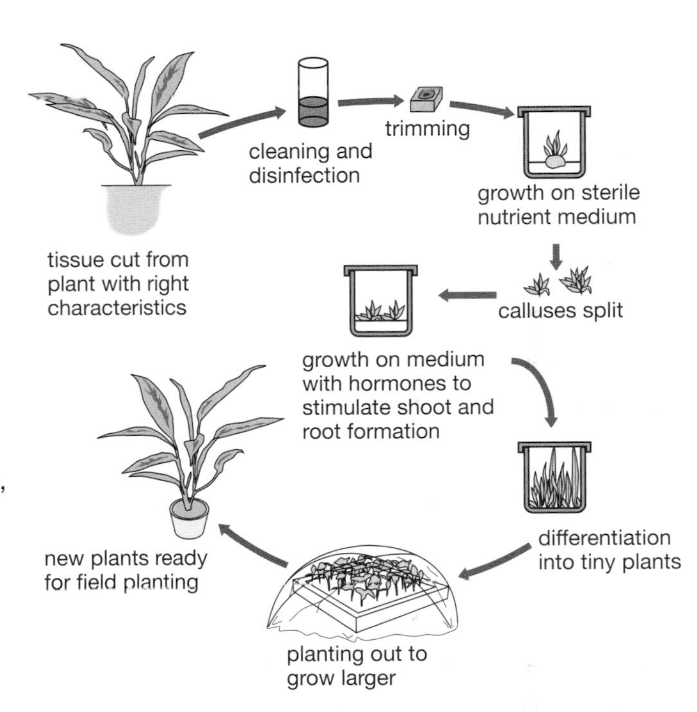

Fig. 5d.01: Micropropagation of banana plants (which do not grow easily from seed)

Micropropagation is fast and very productive, but has high initial costs because of the sterile laboratory conditions and trained staff that are required.

All the plants produced from one parent plant are genetically identical (clones), and the number of new plants grown can be huge. The technique is important commercially as:

- large numbers of plants can be produced for the houseplant market, particularly with unusual colours or markings, or if difficult to grow from seed, such as orchids
- GM plants grow quickly and at a similar rate, as all the plants will contain the additional gene
- virus-free crop plants from species that are important commercially but are difficult to grow from seed, e.g. bananas can be produced. (The use of small numbers of cells in the explant also reduces the risk of viruses in the parent plant being transferred to the new plants.)

Cloning of mammals

Mammals can be cloned using the following technique called adult cell cloning which was used to create Dolly the sheep.

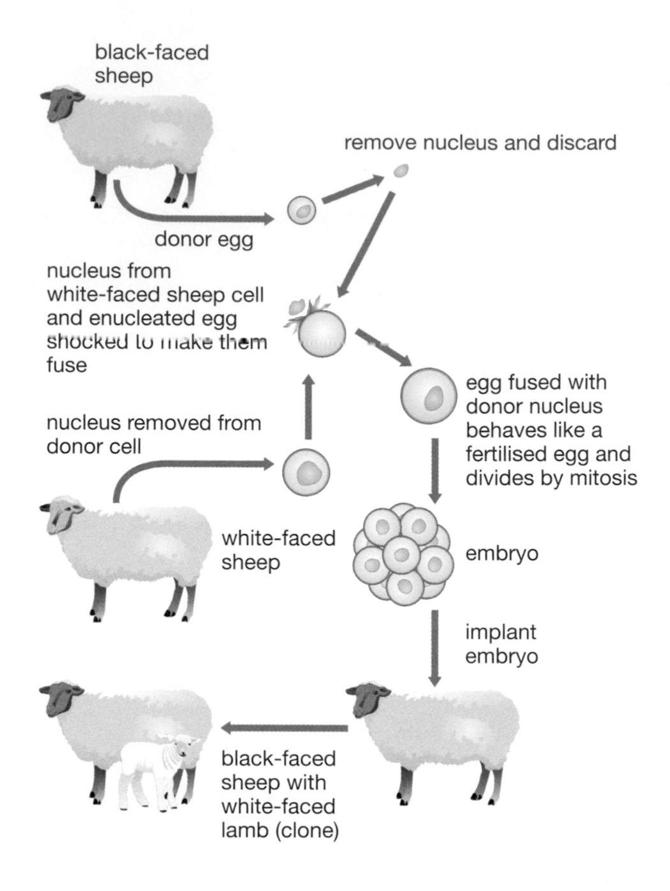

Fig. 5d.02: Making Dolly the sheep – adult cell cloning

- A diploid body cell is taken from the sheep (from the udder, which is easy to sample), and the diploid nucleus is removed.
- The nucleus of a haploid egg cell is removed (the cell is **enucleated**).
- The diploid nucleus is put into the enucleated egg cell.
- The fused cell is given a shock to start it dividing.
- The cell divides to form an embryo.
- The embryo is placed in the uterus of another sheep to develop until it is born.
- The lamb that is born will have the characteristics of the sheep from which the diploid body cell was taken.

Dolly the sheep was the first cloned mammal produced using this method. Since then, the method has been used to produce other cloned mammals such as dogs, rabbits and monkeys.

Embryo transplants are performed by splitting apart the cells of a developing embryo (unspecialised embryonic stem cells). Each unspecialised embryo stem cell divides to form a new embryo. Each new embryo can be transplanted into a new host mother.

Using cloned mammals

Cloned mammals could be used for:

- making commercial quantities of human antibodies – making it cheaper and quicker to produce large amounts for preventing infection
- producing mammals, e.g. pigs, with human-like immune responses so that we can use their organs for transplantation into humans.

The advantages of using transgenic mammals as a source of organs for transplant are:

- there are not sufficient human organs available for transplant for all the patients who need them, and many die before an organ becomes available
- the organs are accepted by the human body without the need to take drugs that suppress the immune system for life which risks infections by microorganisms.

The disadvantages include:

- pigs carry viruses that can be harmful to humans
- it can take many attempts to produce cloned piglets by this method, making it costly
- many people are against the idea of using animals in this way.

You should now be able to:

★ describe how micropropagation is used to produce large numbers of identical plants (see page 173)

★ give an example of how micropropagation is used commercially (see page 173)

★ describe the stages in the production of Dolly the sheep (see page 174)

★ describe the advantages and disadvantages of producing cloned mammals (see above).

Practice questions

1.

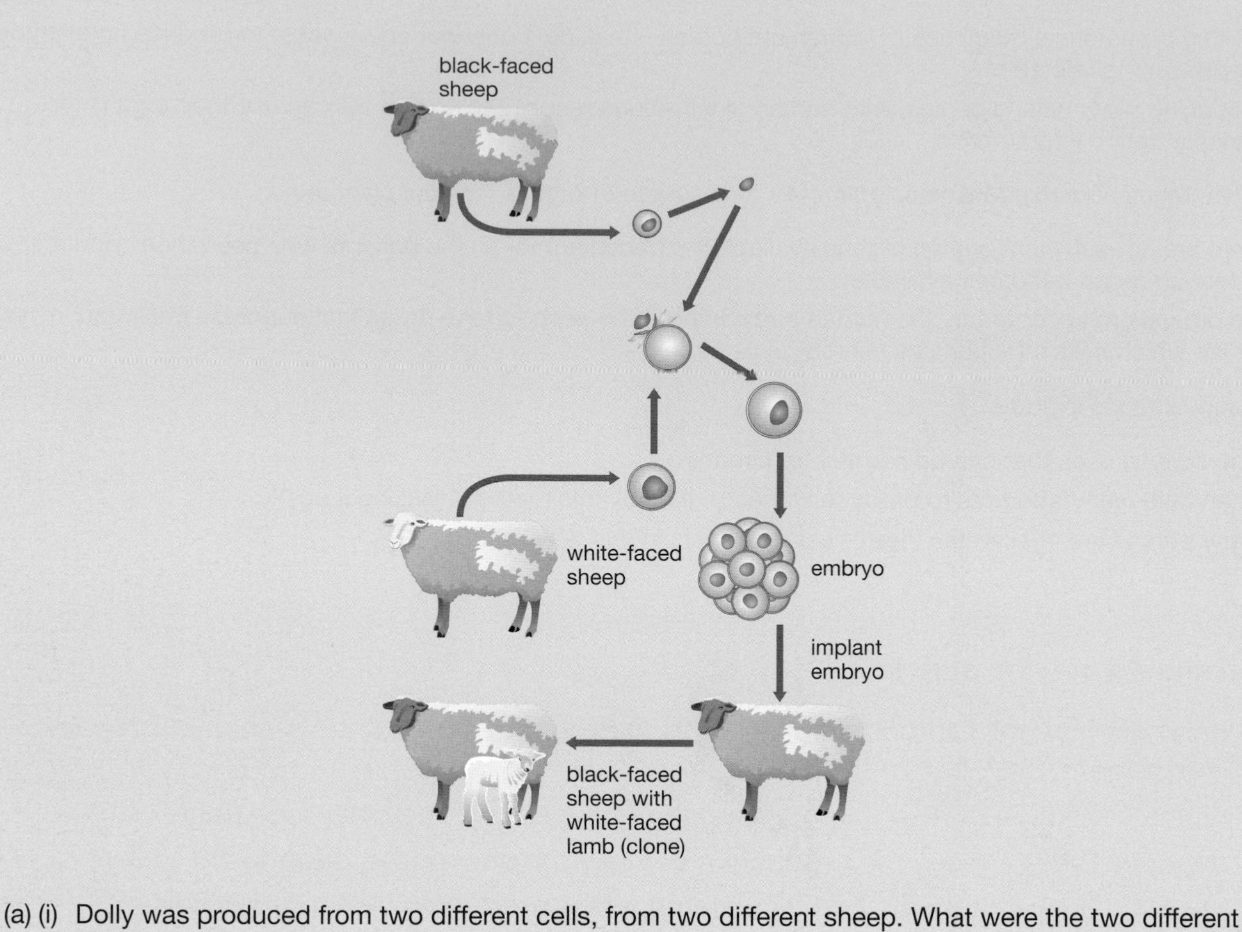

(a) (i) Dolly was produced from two different cells, from two different sheep. What were the two different cells? **(2)**

(ii) What kind of cell division did the spark of electricity start? **(1)**

(iii) Which sheep is Dolly genetically identical to in the diagram? **(1)**

(b) List two ways in which the method which was used to produce Dolly is different from the normal method of reproduction. **(2)**

(c) Suggest one advantage of producing animal clones. **(1)**

2. Micropropagation is used to produce the following crops: bananas, ginger, oil palm, orchids, pineapple, cotton and tea.

(a) Micropropagation is not done in many developing countries because it is so expensive. Explain why it costs so much. **(2)**

(b) Micropropagation is a form of what kind of plant reproduction? **(1)**

(c) Nematodes are parasitic worms that live in the soil, and then infest plants. These nematodes can cause a great deal of damage to young plants. How does micropropagation avoid nematode infestation? **(2)**

(d) What other advantages are there of raising plants by micropropagation? **(3)**

Answers

1 The nature and variety of living organisms

A Characteristics of living organisms

1. (a) Nutrition (absorption of light energy) **(2)**

(b) Response to surroundings (recognition of cold temperature); control of internal conditions (shivering to keep warm) **(2)**

(c) Excretion (carbon dioxide is a waste product) **(2)**

(d) Growth (through cell division); nutrition (obtaining water and nutrients from the soil) **(2)**

(e) Reproduction (through mitosis, a form of asexual reproduction) **(2)**

B Variety of living organisms

1. A tree is classified as a plant because:
- it has chloroplasts for photosynthesis
- its cells have walls made of cellulose
- it stores carbohydrates as starch or sucrose
- it has no nervous coordination and cannot displace itself

A bird is classified as an animal because:
- it has nervous coordination and can move from place to place
- its cells have no cell wall
- it stores carbohydrates as glycogen
- it digests organic material **(8)**

2. Vertebrates have:
- a segmented spine that protects the spinal cord
- a skull that protects the brain
- characteristic internal organs such as kidneys, liver, endocrine organs and a closed circulatory system

All other animals that cannot be classified as vertebrates are invertebrates **(6)**

3. The binomial system uses the genus name and the species name **(2)**

4. Fungi have chitin in their cell walls
Arthropods have chitin in their exoskeleton
Molluscs have chitin in their tongue **(3)**

2 Structures and functions in living organisms

A Levels of organisation

1. At least two explanations must be given

   Correctly chosen organ system and its function (max. 2 marks), e.g. circulatory system, transport oxygen to tissues, carbon dioxide away, soluble products of digestion, hormones, etc.

   Correctly chosen organs and their functions (max. 2 marks), e.g. heart – pumping blood to and from lungs and body; arteries carry oxygenated (except for pulmonary) blood away from heart; veins carry deoxygenated blood to heart (except for pulmonary vein)

   Correctly chosen tissues and their functions (max. 2 marks), e.g. blood – liquid tissue transports substances, clots when stops flowing as in a bleeding wound

   Correctly chosen cells and their functions (max. 2 marks), e.g. red blood cells specialised for the transport and delivery of oxygen; phagocytes – engulf pathogens; lymphocytes secrete antibodies; platelets help form a clot

B Cell structure

1. (a) Twenty onion cells actual size = 0.5 mm **(2)**

   (b) Chloroplasts (very few), vacuoles, cellulose cell wall **(2)**

   (c) Chlorophyll, absorbs light energy **(2)**

   (d) Root hair cells have no chloroplasts and they have a long extension to increase surface area to absorb water and dissolved ions from the soil **(2)**

2. **(1) per row**

	Cell structure	Cell structure's function
1	nucleus	controls activities of cell
2	cytoplasm	where chemical reactions take place
3	cell membrane	controls what is going in and out of the cell

C Biological molecules

	Carbohydrates	Proteins	Lipids
Contains C, H and O	✓	✓	✓
Contains N		✓	
Sometimes contains S		✓	
This group includes enzymes		✓	
Contains glycerol			✓
Subunits are amino acids		✓	
Fungi can digest these	✓	✓	✓
Contains the highest % of high energy C-H bonds			✓
Made at ribosomes		✓	
Made by photosynthesis	✓		

(1) to each row

D Movement of substances into and out of cells

1. (a) Partially permeable **(1)**

   (b) Permeable **(1)**

   (c) Cell membrane **(1)**

   (d) Partially permeable or a cell membrane **(1)**

   (e) Red blood cells burst in a dilute solution because they have filled up with water by osmosis. Haemoglobin is released, solution looks clear **(4)**

   (f) Becomes turgid because of osmosis **(1)**

   (g) Water is lost from potato by osmosis, it becomes flaccid and shorter. Salt has diffused into cells **(3)**

E Nutrition

1. (a) Diffusion, active transport **(2)**

   (b) No. Water is not broken down by digestion. It is absorbed passively by osmosis, then excreted if in excess by the kidneys **(1)**

   (c) Osmosis **(1)**

   (d) Roughage **(1)**

2. (a) Glycerol and fatty acids **(2)**

   (b) Bile emulsifies fats, increases surface area, lipase enzymes work better to digest the lipids **(2)**

   (c) Weight gain/obesity. May cause: heart disease, increased blood pressure; shortness of breath; arthritis; diabetes **(2)**

3. (a) P; B; X; A **(1)**
   (b) Photosynthesis **(1)**
   (c) Respiration **(1)**
   (d) Chlorophyll **(1)**
   (e) Chloroplasts **(1)**
   (f) Magnesium (accept nitrogen as there is nitrogen in the porphyrin ring of the chlorophyll molecule) **(1)**

F Respiration

1. (a) His muscles obtained energy from anaerobic respiration **(2)**
   (b) To repay his oxygen debt **(2)**
   (c) Sugar → ethanol + carbon dioxide = energy **(2)**
   (d) Any suitable answers, such as making bread, brewing alcoholic drinks **(2)**
   (e) Anaerobic respiration **(2)**
   (f) $C_6H_{12}O_6 \rightarrow 2C_3H_6O_3$ (+ energy) **(2)**
   (g) $C_6H_{12}O_6 + 6O_2 \rightarrow 6CO_2 + 6H_2O$ (+ energy) **(2)**

G Gas exchange

1. (a) Mucus-secreting cells; mucus traps bacteria and other particles; ciliated cells move mucus to back of throat **(3)**
   (b) Diaphragm muscle relaxes, diaphragm returns to original domed shape; intercostal muscles relax; rib cage moves back down and in; volume inside lungs decreases and the pressure increases to higher than atmospheric pressure; therefore air is exhaled **(4)**
   (c) **(Two of the following points for 2 marks each; 4 marks available)**

   Smoke contains tar, tar is carcinogenic and chances of developing cancer increases

   Ciliated cells stop working, mucus containing bacteria (and tar) sinks into bronchi, bronchioles, alveoli and the chances of chest infections increases

   Coughing increases, alveoli become damaged, emphysema

   Carbon monoxide reduces oxygen in blood, leads to low birth weight babies in pregnant smokers; heart disease in smokers as heart works harder to pump blood around

   (d) Trachea, bronchi and bronchioles **(3)**
   (e) Large surface area; good blood supply; thin walls **(3)**

2. (a) Yes **(1)**
   (b) Carbon dioxide **(1)**
   (c) Diffusion **(1)**
   (d) Guard cells (do not accept stomata – they are openings and are controlled by the guard cells) **(1)**
   (e) Labelled diagram of four boiling tubes with hydrogen carbonate indicator. Two tubes with *Elodea*. One *Elodea* tube and one tube with indicator only are kept in the dark, the other pair of tubes in the light **(5)**

H Transport

1. (a) 28 g **(1)**

   (b) 29 g **(2)**

   (c) Water moves up a plant by transpiration and most is lost from the stomata. Most of the water a plant absorbs is to maintain the shape of the plant **(4)**

   (d) Some water stays in the cells to keep them turgid. Some water is used in photosynthesis **(1 way for 1 mark)**

   (e) The conditions may vary – humidity, temperature, wind speed, all of which affect the rate of transpiration **(2)**

   (f) The cells would not be turgid; the plant would wilt **(2)**

2. (a) Vena cava empties deoxygenated blood into right atrium. Right atrium contracts, pushing blood past atrioventricular valves into relaxed right ventricle. Semilunar valves of pulmonary artery shut. Right ventricle contracts, pushing blood out through pulmonary artery, past semilunar valves. Atrioventricular valve snaps shut **(5)**

   (b) Coronary artery(ies) become blocked by blood clot or plaque; oxygen supply to contracting heart muscle reduced; heart muscle stops contracting **(2)**

   (c) Carbon monoxide inhaled in tobacco smoke combines with haemoglobin and stops haemoglobin from carrying as much oxygen as it should. Respiring tissues do not receive as much oxygen, heart pumps harder and faster; this leads to damaged, blocked arteries

   Also long-term effect – emphysema – reduced gas exchange and oxygen absorption, heart works harder, arteries become damaged **(3)**

   (d) Diet rich in saturated fats, refined sugars and salt can lead to obesity and high blood pressure. Heart has to work harder to carry oxygen to tissues. Arteries are damaged, become blocked **(2)**

   (e) Oxygen in haemoglobin of red blood cells in capillaries surrounding alveoli in lungs enters pulmonary vein at low pressure. Pulmonary vein empties into left atrium. Left atrium contracts, pushing blood past atrioventricular valves. Blood enters relaxed left ventricle, and semilunar valves of aorta shut. Left ventricle contracts, closing atrioventricular valves and pushing semilunar valves open. Oxygenated blood goes into aorta at high pressure, and on to other arteries. Arteries lead to capillaries where gas exchange occurs. Oxygen diffuses from haemoglobin in red blood cell to respiring tissues **(5)**

I Excretion

1. (a) A – Bowman's capsule; B – glomerulus; C – collecting duct **(3)**

   (b) Glucose, water, ions, amino acids, glycerol, fatty acids, urea **(name 5 for 5 marks)**

   (c) Small molecules, soluble molecules **(2)**

   (d) Glucose, amino acids, glycerol, fatty acids, required ions, water if required **(name 4 for 4 marks)**

   (e) Water, ions, urea **(3)**

   (f) Bladder **(1)**

2. (a) During dialysis, excess water is removed because of osmosis and excess ions diffuse out into dialysis fluid **(2)**

(b) Kidney transplant is permanent so lifestyle improves and is cheaper in the long run **(2)**

3. (a) Kidney tubule removes urea and reabsorbs water, so urea concentration will rise **(1)**

(b) Hormone **(1)**

(c) Pituitary **(1)**

(d) Low *blood* water levels **(1)**

(e) Less ADH secreted when blood water levels are higher than normal. Collecting tubules reabsorb *less* water. Larger volume of more dilute urine is produced **(3)**

J Coordination and response

1. (a) Shoot tip **(1)**

(b) When the tip was covered, no response to the light; the other two where tip is exposed, response to light occurs; when base of shoot covered, response to light occurs **(2)**

(c) Auxin is secreted from meristem tissue at shoot tip and diffuses down to all of the shoot. In cells exposed to light in shoot, auxin has no effect on increased growth (cell division). In cells in shade, auxin increases cell division **(3)**

(d) Shoot grows towards light source, maximising light absorption for photosynthesis **(2)**

2. (a) X – sclera; Y – retina; Z – iris **(3)**

(b) Fovea at back of eye, little notch just above optic nerve **(1)**

(c) Cornea and lens **(2)**

(d) If optic nerve cut, nerve impulses from rods and cones would not be sent via sensory neurones to brain. **(2)**

(e) Looking out of the window (relaxing your eyes – daydreaming!) the ciliary muscles relax, the suspensory ligaments become taut, the lens flattens. Looking back at the question (studying – hard work) the ciliary muscles have to contract. Suspensory ligaments slacken, lens returns to naturally round shape for near focusing **(4)**

3. (a) (i) Message is transmitted by chemicals called hormones, released in blood **(1)**

(ii) Speed of transmission is slower than along neurones **(1)**

(iii) Effects of hormones are longer lasting compared to neurones **(1)**

(b) Sensory neurone, relay neurone in CNS, motor neurone, effector **(4)**

(c) Where endings of two neurones meet. Nerve impulse is transmitted across synapse when a chemical is released and diffuses across. Only one neurone secretes the chemical, so nerve impulse only goes in one direction **(3)**

3 Reproduction and inheritance

A Reproduction

1. (a) X – petals; Y – sepals; petals attract insect pollinators; sepals protect the flower bud as it develops **(2)**

   (b) (i) Anthers **(1)**; (ii) Stigma **(1)**

   (c) Grows a pollen tube down style to ovule **(2)**

   (d) Pollen nuclei first formed by meiosis **(1)**

   (e) Mitosis **(1)**

2. (a) Menstruation **(1)**

   (b) The uterus lining grows again due to oestrogen and progesterone in readiness for a fertilised egg **(1)**

   (c) Should repeat the cycle shown on the diagram. The thickness of the uterus lining increases and is maintained briefly in case a fertilised egg implants **(3)**

   (d) Meiosis **(1)**

   (e) Mitosis **(1)**

   (f) Her eggs could be removed surgically using a fine tube inserted into her abdomen **(2)**

   (g) (i) Eggs mixed with sperm in petri dish to increase chances of fertilisation, to check that fertilisation has taken place so only fertilised embryos are put back in the woman **(2)**

   (ii) To mimic the natural time before an embryo would reach the uterus after fertilisation in the oviduct, and to ensure implantation into the uterine lining **(1)**

   (h) The fertilised egg is implanted in the woman's uterus, where the baby develops. The baby does not develop in a test tube **(2)**

   (i) Placenta, umbilical cord, amniotic fluid, increased blood supply to uterus, increase in size **(3 changes for 3 marks)**

B Inheritance

1. (a) XY chromosomes **(1)**

   (b) Instead of a pair of chromosomes 21, there are three – there is an extra chromosome 21 **(1)**

   (c) The 21st pair of chromosomes did not separate as is normal for chromosome pairs during meiosis. At fertilisation, the chromosome 21 from the sperm means that there are three chromosomes 21 **(2)**

(d) **(4)**

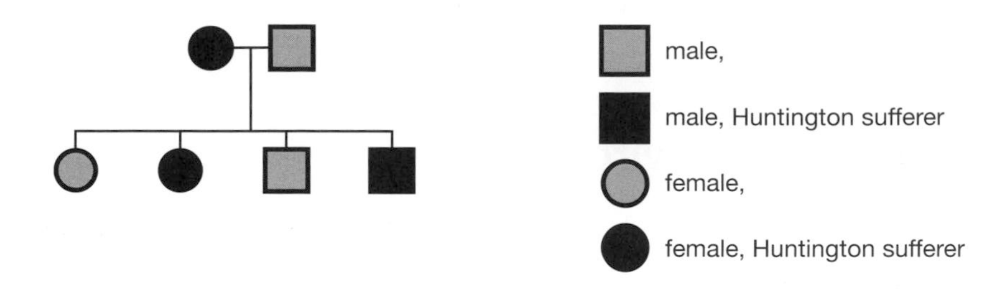

(e) **(4)**

	H	h
h	Hh	hh
h	Hh	hh

(f) A disease that is inherited from your parents and which is incurable **(2)**

2. Genetic variation existed in the German cockroach population, so some individuals had the allele for disliking sugar. Cockroach traps were introduced, with an insecticide in glucose as bait. Cockroaches that ate the insecticide died. Cockroaches that did not like glucose did not eat the insecticide and glucose, so they survived. They reproduced, and passed their glucose-hating alleles on to the next generation **(5)**

4 Ecology and the environment

A The organism in the environment

1. (a) B **(1)**

   (b) D **(1)**

   (c) C **(1)**

   (d) Amount of glucose available, amount of oxygen available, amount of amino acids available, the amount of toxic waste products building up **(3 factors for 3 marks)**

   (e) Line should continue down to level of A or lower **(e)**

   (f) Above nutrients and oxygen were used up, respiration rates and growth rates slowed down, and finally the build up of toxic waste products killed all the bacteria **(1)**

2. (a) Cod had been overfished and too few cod were left to reproduce **(2)**

   (b) Cod numbers should recover if fishing is banned **(1)**; the fishermen and industries based around the fishing communities will not make money from fishing, and will have to diversify **(1)**

   (c) More efficient fishing techniques were used; more fishing boats; increased demand for fish due to human population growth (baby boom after WWII) **(1 reason for 1 mark)**

   (d) With fewer cod, lobsters had fewer predators eating them. More lobsters survived and reproduced **(2)**

B Feeding relationships

1. (a) Photosynthesis **(1)**

   (b) Light energy is changed to chemical energy **(1)**

   (c) Excretion, respiration, undigested food, heat **(3 ways for 3 marks)**

   (d) Trophic level or energy flow **(1)**

   (e) Seal populations will drop as increased competition for prey. **(2)**

2. (a) $\frac{330 \text{ kJ} \times 7\text{days}}{400 \text{ kJ}} = 5.8$, so 6 stoats a week is the $\frac{\text{energy requirement of the owl}}{\text{energy stored in a stoat}}$ **(1)**

   (b) $\frac{20 \text{ kJ}}{12 \text{ kJ}} = 1.6$ oat plants a day is the $\frac{\text{energy requirement of a mouse}}{\text{energy stored in grass}}$ **(1)**

   (c) $\frac{80 \text{ kJ}}{100 \text{ kJ}} = 0.8$ or 1 mouse is the $\frac{\text{energy requirement of a stoat}}{\text{energy stored in a mouse}}$ **(1)**

   (d) Pupils must remember to multiply values by 7 to calculate this, then by numbers of animals eaten:

   5.8 stoats eaten by 1 owl

   4 × 7 = 28 field mice eaten for 1 stoat, therefore 28 × 5.8 = 162 mice

   3.6 × 7 = 25.5 oat plants a week for 1 field mouse, therefore 25.5 × 28 = 714 oat plants **(4)**

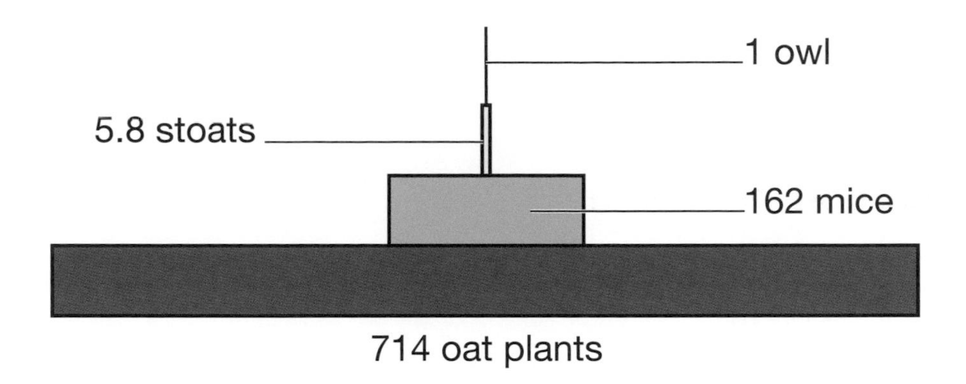

C Cycles within ecosystems

1. (a) The bacteria suddenly have a sugar food source for respiration at the outflow. Quickly their numbers increase. As the sugar is used up, the rate of multiplying of bacteria slows down **(2)**

   (b) Oxygen demand increases **(2)**

   (c) Numbers of fish decrease downstream, then start to increase the further away from the outflow **(2)**

   (d) By photosynthesis of aquatic plants; by diffusion from the air **(2)**

2. (a) Photosynthesis **(2)**

   (b) Respiration **(2)**

   (c) Eat them **(2)**

   (d) (i) If the coal is burnt **(2)**

   (ii) Increased greenhouse effect **(2)**

   (iii) Decay/decomposition **(2)**

D Human influences on the environment

1. (a) To clear land for humans to live; logging for timber; clearing land for roads; clearing land for crops and cattle **(3 reasons for 3 marks)**

   (b) (i) They are going up **(1)**

   (ii) Burning trees releases carbon dioxide, and fewer trees means less photosynthesis **(2)**

   (iii) Increased greenhouse effect; less CO_2 is absorbed for photosynthesis; wood burning gives off more CO_2; loss of trees means fewer leaves falling – less cycling of nutrients **(2)**

   (c) National Parks set up; captive breeding programme and release **(2)**

2. (a) Microorganisms cause decay. They need moisture to live. **(1)**

   (b) Microorganisms need oxygen to respire and to make the enzymes that bring about decay. **(2)**

   (c) By aerating it. Microorganisms respire and release heat **(2)**

   (d) Microorganisms have broken down large insoluble molecules into small soluble molecules that plants can absorb through their roots. Nitrates are vital for proteins, magnesium for chlorophyll, phosphates for DNA (cell division) **(3)**

5 Use of biological resources

A Food production

1. (a) Cane toad numbers increased **(1)**

   (b) Decreased/became extinct **(1)**

   (c) No chemicals used so usually less damage to environment; pest population reduced to non-damaging level, so local food webs not as affected; very useful in restricted environments such as glasshouses **(3)**

   (d) Advantages: reliable effect; can be used when needed; applied directly to the crops **(2 advantages for 2 marks)**

   Disadvantages: many made from non-renewable oil; stable chemicals may last a long time in the environment; may affect species other than the pest, damaging food webs; may cause health problems for farmers and other local people; insects may develop resistance to pesticide **(2 disadvantages for 2 marks)**

2. (a) Fungi **(1)**

   (b) (i) Diffuser – sends bubbles of air through the mixture in the fermenter **(1)**

   (ii) Water-cooled jacket – removes excess heat from respiring microorganisms **(1)**

   (iii) pH probe is for monitoring pH so pH can be adjusted **(1)**

   (c) To prevent contamination from other microorganisms **(1)**

   (d) Penicillium **(1)**

   (e) The fermenters can be built anywhere that food is needed; the nutrients for growing the fungus can use waste substances from other processes, such as waste straw from crop plants; more protein is produced more quickly and cheaply than growing animals, e.g. beef cattle; single cell protein contains more fibre and little fat, making it healthier than eating meat; because of the high fibre content, it has the texture of meat; it has a pleasant taste **(3 advantages for 3 marks)**

B Selective breeding

1. (a) Genus name **(1)**

   (b) Artificial selection **(1)**

   (c) Isolate a population of mosquitoes; record which offspring mature and reproduce quickest; isolate these individuals; crossbreed them; continue to select the shortest life cycle mosquitoes and crossbreed them for many generations until all offspring have two-week life cycles **(4)**

   (d) So the malaria parasite does not have long enough to develop in the mosquito, and therefore the mosquito cannot transmit it **(2)**

   (e) They want the shorter life cycle mosquitoes to be able to fly and find mates and spread their alleles **(2)**

   (f) Production of large numbers of eggs so the shorter life cycle mosquito numbers increase in the population **(2)**

2. (a) Thick coats; high milk yield; high reproduction rate; ability to eat low quality food during the winter; disease resistance; tame; ability to absorb oxygen efficiently **(4 characteristics for 4 marks)**

(b) Respiration **(1)**

(c) Faster growth rate; higher percentage muscle tissue **(2)**

(d) Swiss cows all had low genetic variation and few had alleles for resistance to new disease **(2)**

C Genetic modification

1. (a) Monera **(1)**

(b) Bacteria have no nuclei and only ribosomes and no other organelles; bacteria have plasmids; bacteria have only one chromosome; bacterial chromosome is a circle **(3)**

(c) Plasmids are small rings of DNA and often have genes for antibiotic resistance **(2)**

(d) Restriction enzymes cut out human insulin gene; these enzymes also cut open the plasmid DNA; human insulin gene then joined to open plasmid by ligase enzymes **(4)**

(e) Asexual reproduction **(1)**

(f) Diabetes **(1)**

(g) Insulin from other animals can cause immune reactions in humans; human patient therefore must stop taking that animal's insulin; insulin from other animals may not be pure and may introduce a disease to the patient **(2)**

2. (a) Nucleus **(1)**

(b) Chromosome **(1)**

(c) Fewer fossil fuels burnt to make nitrate fertilisers, less carbon dioxide emissions; also less nitrate fertilisers means less eutrophication of streams and rivers, biodiversity in rivers and streams stays high. Also give marks if eutrophication described in detail **(4)**

D Cloning

1. (a) (i) Udder cell (mammary gland cell) and an egg cell **(2)**

(ii) Mitosis **(1)**

(iii) Dolly is identical to sheep from which udder cell nucleus was removed **(1)**

(b) Only one gamete used; no sexual intercourse; only one genetic parent; no surrogate mother **(2 ways for 2 marks)**

(c) Can make large quantities of pure human protein product; mammals live a long time, so quality control is easier compared to genetically engineered bacteria; can make many clones from one embryo **(1 advantage for 1 mark)**

2. (a) You need highly trained staff and high tech laboratories **(2)**

   (b) Asexual reproduction **(1)**

   (c) Because uninfected tissue is used, then it is sterilised and kept in sterile conditions as it grows **(2)**

   (d) To produce large numbers of plants for the houseplant market, particularly with unusual colours or markings, or if difficult to grow from seed such as orchids; to produce large numbers of GM plants quickly, as all the plants will be genetically identical and therefore contain the GM characteristic; to produce large numbers of virus-free crop plants from species that are important commercially but are difficult to grow from seed, e.g. bananas (the use of small numbers of cells in the explant reduces the risk of viruses in the parent plant being transferred to the new plants) **(3)**